Performance Automotive

Engine Math

John Baechtel

S-A DESIGN

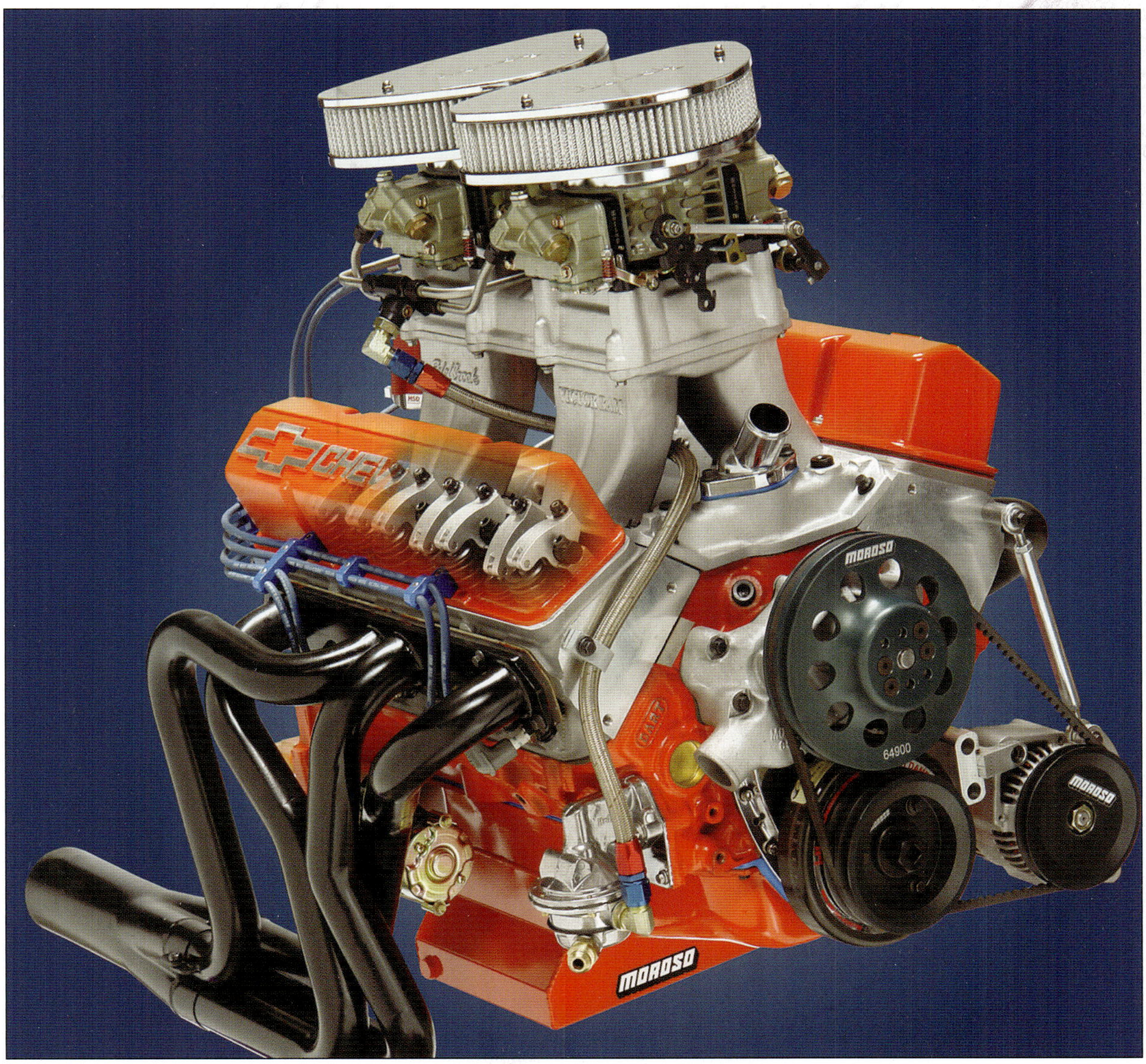

CarTech®

CarTech®

CarTech®, Inc.

6118 Main Street, North Branch, MN 55056
Phone: 651-277-1200 or 800-551-4754
Fax: 651-277-1203
www.cartechbooks.com

Edit by Scott Parkhurst
Layout by John Baechtel

ISBN 978-1-934709-47-4
Item No. SA204

Library of Congress Cataloging-in-Publication Data

Baechtel, John.

 Performance automotive engine math / John Baechtel.
   p. cm.
 ISBN 978-1-934709-47-4
 1. Automobiles--Motors--Mathematics. 2. Automobiles--Motors--Mathematical models. I. Title.

 TL210.B284 2011
 629.25001'51--dc22

                          2010042285

Written, edited, and designed in the U.S.A.
Printed in China
10 9 8

Title Page: This first-gen small-block Chevy represents a significant investment from its owner, but why spend big money to get all the best-quality parts if they're not going to work together as a system? By researching the mathematical capabilities of the various components, and designing the engine as a collection of complementary subsystems, the final result is a super-efficient powerplant that delivers the most-possible power very efficiently.

Back Cover Photos

Top Left: Here a complete header design has been modeled on a V-8 engine already installed in a chassis. Note the starter blocks and handwritten notes on each segment indicating the lengths and degree of alignment so the header fabricator can easily duplicate the finished model in metal.

Top Right: Degreeing the cam requires accurately locating TDC with a degree wheel and pointer and a dial indicator set up to read lifter travel as you rotate the engine.

Bottom Left: Determine combustion chamber volume by filling the chamber with water or alcohol from a graduated burette calibrated in cubic centimeters (cc's). Tighten a spark plug in the chamber with both valves installed. Then use a light grease to seal the deck surface. Place the plastic cc plate over the chamber and position the head so the fill hole is at the highest point. Fill the chamber and read the cc's off the burette. Divide by 16.4 to convert to cubic inches.

Bottom Right: To calculate the average intake runner cross-section measure the height and width of the runner entry and exit with dial calipers. The entry and exit are usually different so take the average of the two cross-sections.

DISTRIBUTION BY:

Europe
PGUK
63 Hatton Garden
London EC1N 8LE, England
Phone: 020 7061 1980 • Fax: 020 7242 3725
www.pguk.co.uk

Australia
Renniks Publications Ltd.
3/37-39 Green Street
Banksmeadow, NSW 2109, Australia
Phone: 2 9695 7055 • Fax: 2 9695 7355
www.renniks.com

Canada
Login Canada
300 Saulteaux Crescent
Winnipeg, MB R3J-3T2 Canada
Phone: 800 665 1148 • Fax: 800 665 0103
www.lb.ca

ACKNOWLEDGMENTS

Books like this rarely happen in a vacuum. This one is the result of 40 years of practical application and the good fortune of lengthy association with brilliant individuals whose willingness to share their knowledge is surpassed only by their passion for engines. Most of the contributors hold advanced degrees and have excelled in their respective fields. Most of the formulas herein are widely known. A few are recalled from a discussion of automotive math formulas compiled by Larry Atherton in the mid 1990s while working on the original version of Motion Software's DeskTop Dyno software, and I have seen them elsewhere from time to time. Larry polled automotive engineers deep within the bowels of GM and Ford engine labs and, while most of what he learned is beyond the scope of this book, it contributed greatly to my further understanding of the complicated details of internal combustion engines and it is still available in Motion Software's extensive online manuals for DeskTop Dyno and Dynomation. Among the many formulas I recall learning from Atherton's early investigations is one for the instantaneous acceleration of the piston from TDC, which I had seen nowhere else until it was published in Forbes Aird's excellent *Automotive Math Handbook*.

A little-known formula for calculating torque peak location and exhaust header primary tube size was contributed by my mentor extraordinaire and perhaps the most widely respected authority in our industry, Jim McFarland. Jim also afforded me the honor of checking my work for mathematical errors and penning the Foreword for the book. I should also acknowledge John Lawlor, whose groundbreaking *Auto Math Handbook* has improved the knowledge of countless engine enthusiasts for decades. Another significant contributor is Jeff Smith from *Car Craft* magazine, a valued colleague whose 30-year friendship has contributed enormously to my ever-expanding automotive database, and who generously consented to giving the manuscript a read for accuracy. Thanks also to Sigrid Michelsen and my best friend Larry Beatson for their contributions to the initial research for the book. They helped immensely with supporting details such as the origins of pi, conversion specs, and the history of complicated mathematical equations.

Many thanks also to my daughter Tamara, a graphic artist and web designer, who helped immensely with various layout issues during production of the book. Her tireless efforts helped ensure the look of the final product, and helped bring it in on time. And, of course, thanks and acknowledgements to my old Bonneville racing partner, Chuck Jenckes, a NASCAR and Formula 1 engineer and endless source of performance engine knowledge.

Finally, my unending thanks to Publisher Dave Arnold, whose patience and generosity remain unsurpassed. Likewise to Scott Parkhurst, who had my back throughout the process, shot many of the photos, and whose ability to edit a pile of gibberish into a cohesive and useful package never ceases to amaze me.

— *John Baechtel*

John Baechtel is a former editor of Car Craft *and* Hot Rod *magazines and was the founding partner of the Westech Performance Group engine dyno testing facility. He is also a member of the Bonneville 200 MPH club, holding both SCTA and FIA International speed records. He drove the first production-based Mustang Cobra past the 200-mph mark in 1993 and was the driving force behind the restoration of the Summers Brothers Land Speed Record car for The Henry Ford museum. With more than 35 years of high-performance engine and vehicle testing under his belt, he is still passionate about engines and high-performance technology. He currently serves as a technical consultant to several performance aftermarket companies while pursuing his interest in land speed racing and grooming his extensive collection of land speed record model cars and memorabilia.*

FOREWORD

It was during the time I was herding the R&D cats at Edelbrock that I met John Baechtel. He was a staffer at *Car Craft* magazine and had ventured into the bowels of Edelbrock's renowned engine dyno facility for one of many such visits he would pay the company. I recall two distinct traits that became among his trademarks: he asked good questions, and, unlike many of his journalism contemporaries, he also listened to the answers. It was not only refreshing, it also spawned the company's willingness to embrace what became a string of editorial efforts John crafted over the ensuing years as our friendship matured.

In fact, his approach to editorial activities drew the attention of Petersen Publishing Company management, leading to him becoming the executive editor of *Hot Rod* magazine, the company's flagship publication, perhaps the first benchmark in his then-emerging career. After a 10-year stint at *Hot Rod*, he returned to *Car Craft,* obviously having demonstrated his editorial skills had risen to a level of responsibility qualifying him to assume responsibility for *Car Craft's* entire editorial content.

During the course of his growth with Petersen, and common for aspiring journalists in an international magazine of *Hot Rod's* stature, John began to be noticed by other automotive communities. Pontiac Motorsports chose him to author their factory performance manual touting the Division's IMSA "GTP Lights" car and its technical highlights. At this point, although contributing to the overall editorial tone of the magazine, John's leaning was more to the technical side of the equation. He was becoming more than a magazine writer and reached out into other areas of his budding interests. Notably, he became qualified to drive on the Bonneville Salt Flats, displaying these skills by the building and piloting a 1991 Pontiac Firebird to an FIA record speed of 221.511 and a C/Gas Coupe record. This further grounded him in future development of the land speed interests that remain with him today.

Possibly out of his desire to dig deeper into the technical side of performance cars and what makes them tick, he departed the ranks of journalism and became the driving force behind the creation of the Westech Performance Group, a West Coast-based R&D shop serving the specialty parts industry, automotive OEM, and journalism communities. It was in this environment that his peers really learned how John was able to culminate his range of prior automotive experiences with a keen desire to help others explore, resolve and achieve a wide variety of performance-engine-related objectives.

To put all this into perspective, it has been by the culmination of events that allowed John to help identify and quantify the means required for understanding how performance and racing engines work that he included the necessary mathematical tools in his repertoire of shared information. Mathematical processes and the building of solid engine packages are inseparable if success is to be achieved. It is on that basis that this book is founded. Many years of practical and hands-on experience are woven into its editorial fabric addressing automotive math. Having known John for more than 30 years and observed his development and record of performance in that time, it is easy to commend the readers of this book to the results of all his hard work.

— *Jim McFarland*

Jim McFarland is a former editor of Hot Rod *magazine, former chief engineer at Edelbrock Corporation during the heyday of high-performance intake manifold development and one of the most broadly respected leaders in the high-performance industry. Upon leaving Edelbrock, Jim formed his own high-tech R&D lab and think tank serving elite corporate clients such as General Motors. He currently enjoys a more relaxed pace as a consulting engineer for SEMA and some of the top performance companies within its elite membership.*

INTRODUCTION

"Do the math" is easy for some people to say, but for many of us the mere mention of math provokes acute anxiety and the urge to flee. Some of us have barely mastered the fine art of counting change and even fewer of us attempt to manage a balanced checkbook. So what qualifies me to write a math book? Very little, I'm afraid. I was a poor math student in college and I have always had the same innate fear of math that many of us feel. Most of us prefer having root canal work over even the simplest of equations. Yet almost every facet of our lives can be reduced to mathematics; most of it functioning quietly in the background to make life easier. Through the simple expedient of repetition we eventually learn just enough math to support our basic needs. Repetition and familiarity are the keys and those are my only qualifications.

For the first 17 years of my career I was technical editor for *Car Craft* magazine, then executive editor of *Hot Rod* magazine and finally back to *Car Craft* as editor for several years. I spent most of my time building and testing engines and cars at other people's facilities. Little by little, the necessary math became second nature. When I finally opened my own dyno testing lab in the mid 1990s, I had to polish my math skills even further to make certain that I knew and understood more than, or at least as much as, the most savvy engine builders and customers who used my services. One thing I learned early is that there is always a source that can be referenced to solve a problem.

That is what this book is for. Somebody smarter than the rest of us figured it all out. We're just looking for the basics and the practical application. This is a reference book; a source, if you will. Thanks to the real

mathematicians and engineers, wherever they are, we can all use these basic equations to solve problems in our projects. My goal is to help you understand the practical application through repetition. This same "no pain" method can be used by any performance enthusiast to learn and apply the basic math that makes the pistons go up and down and the wheels go round and round. If you have the right equations and learn to work them through repetitive use, your engine math butterflies will quickly vanish.

Please note that we are actually working with math equations, not formulas. For our purposes this is pretty much mathematical nitpicking, but any time you have

A scientific calculator includes a square root key, a cube root key, and a key for squaring numbers. These keys and others allow you to quickly make any calculation in this book.

an equal sign between any combination of known and unknown values, you have an equation. By entering the known variables and using algebraic manipulation you can solve the equation. Because both sides always have to remain equal, you can move things back and forth (transposing), or add, subtract, multiply, and divide equally on both sides to keep the equation balanced. In most cases the equations are going to have the unknown we are seeking on the left side and a bunch of variables on the right that, when properly manipulated, yield the answer. Since I'm spelling it all out for you, all you have to do is work the basic arithmetic. If you see brackets [ ] or parentheses ( ), perform the operations inside of them first, almost like a separate problem. Multiply and divide first and then add and subtract to simplify operations.

Now even though we are working with mathematic equations, we are still going to follow popular convention and call them formulas because that is what car people are most familiar with. When all is said and done, it won't matter.

In addition to all the basic engine math formulas, I have included practical examples and as much background reference material as possible. That includes basic formulas for finding the area of squares, circles, and triangles, and the circumference of a circle using pi. It also includes explanations of different units, linearity, exponents and powers, rates and speeds, weights and measures, pressures and temperatures and so on. Some of the core principles discussed include leverage and fulcrums, pressure and vacuum, motion and inertia, elasticity, friction, density altitude, and other useful information.

The formulas require nothing more than basic algebra and square roots; nothing too fancy or over your head. Most can be worked on a simple household calculator; but do yourself a favor and pick up an inexpensive scientific calculator that includes separate keys for pi, square root, and parenthesis. You can just ignore the rest of the scientific functions. I've pointed out calculator key pad sequences for many of the formulas so they're easy to follow.

You'll also find descriptions of all the tools you will be using to take your measurements and make your calculations. You'll find many of them indispensable and

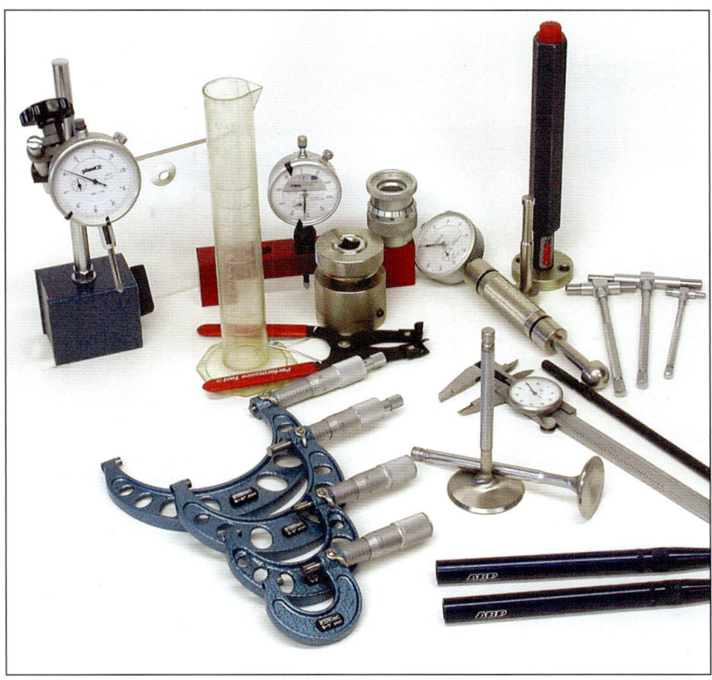

Precision measurement tools are required to obtain accurate results when blueprinting an engine. These include outside micrometers, digital calipers, dial indicators, snap gauges, height micrometers, graduated burettes and more.

inexpensive. Once you become familiar with them, most of your measurements and calculations become second nature. For computer buffs, Chapter 13 describes how to enter all the formulas into a simple spreadsheet, such as Microsoft Excel or any of the open document spreadsheets now available online. You can build your own personalized engine math page where you can enter values and see instant results that you can print for future reference.

The technology available to all of us is increasingly simple and easy to use. I sincerely hope that this book will help you take advantage of it for all your engine math needs.

It's also worth noting that while mathematics is a very rigid discipline, concepts and formulas are rarely set in stone. Over time, theory and corresponding calculations have a way of evolving. It is relatively unlikely that you'll encounter many changes from what you see in this book, but keep an open mind and be willing to examine and evaluate new theory and math as it may emerge over time.

Basic Math and Science

You can't do engine math without numbers. The numbers we use are measured with precision tools or, in some cases, they are assigned values for the purpose of brainstorming proposed modifications and theoretical results. Measurements are taken in what we all recognize as U.S. Customary units of measure such as feet, inches, pounds, and gallons; international SI or metric units such as millimeters, centimeters, and liters are also relevant because many new cars use a combination of both. Once dimensions and other measurements are recorded they can be mathematically manipulated to tell us almost anything we want to know about our engines, including how they might perform and how we might modify them.

Engine math deals with rod and stroke lengths, bearing diameters and clearances, cylinder volumes, bore/stroke ratios, piston weights and speeds, cylinder pressures, atmospheric temperature, and so on.

- Volumes are calculated in cubic inches, cubic centimeters, or liters (appropriate conversion factors are available).
- Engine, piston, and bearing speeds are a function of time so they are measured in feet per minute (ft/min) or feet per second (ft/sec).
- Atmospheric temperature measurements are noted in degrees Fahrenheit.
- Pressure measurements are given in pounds per square inch (psi).

Core Principles

You are probably familiar with the core principles taught in high school science classes. Not surprisingly, all of them apply to engine projects in one way or another. Basic physics affects every function inside your engine and because most of the principles are straightforward and easy to visualize, it's easy to understand how they affect engine performance.

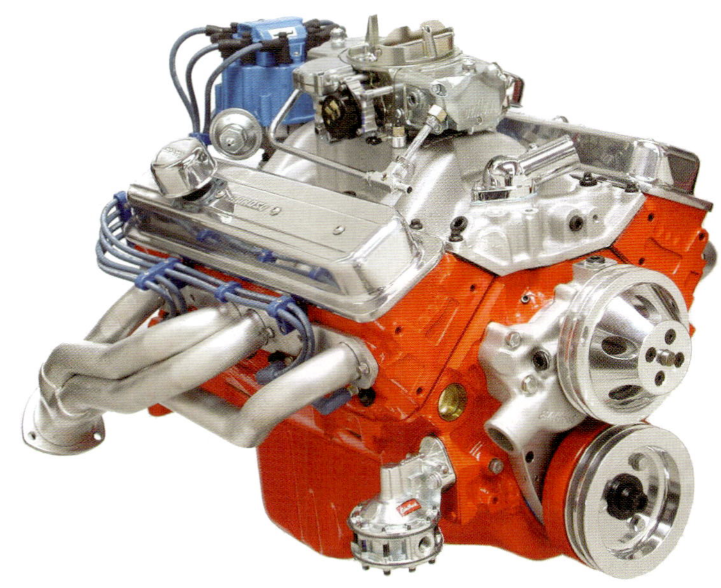

The math going on inside your performance engine is easy to grasp once you understand the formulas and how they relate to engine performance.

Before I dive into the heart of the engine to work some basic short-block math, let's take a brief moment to review some of the core principles that affect engine operation. As you progress through the book, you'll find that each of the following principles apply to engines in one or more ways.

Acceleration

The rate of increase of speed or velocity. In engine math it is usually noted in feet per second (ft/sec) or feet per minute (ft/min). It represents a change in velocity from a state of rest (no motion or speed) to a given speed, or from a given speed to a greater speed over time. Deceleration is a measure of change in velocity from a given speed to a lesser speed or to zero speed.

Area

The extent or measure of a plane surface or the surface of a solid object. Area is represented in square inches or square centimeters. In engine math it might represent the surface of a piston top or the cross-sectional area of an imaginary plane sliced through an intake port at 90 degrees to the direction of flow. It may also represent the open space of a throttle bore through which air passes, commonly called throttle area.

Circumference

The length of a boundary line of a given circle or the perimeter of a closed curve bounding a plane area. Cylinder bores and throttle bores have a fixed circumference which can be converted to an area measurement by squaring the diameter of the given circle and multiplying by the constant 0.7854. Or multiply 2 x pi x the radius of the circle.

Density

The quantity of something by unit of measure or the mass per unit volume of a substance under specified conditions of pressure and temperature. This is a measure of how tightly packed the atoms and molecules are within any given space—weight divided by volume and noted in pounds per cubic inch squared ($\text{lbs/in}^3)^2$.

All the core elements of math and science are present inside every engine. This view of a supercharged Chevrolet Corvette engine reveals the complexity of component relationships that are governed by basic mathematical principles. (Courtesy Chevrolet Motor Division)

Differential

Showing an amount or degree of difference between two quantities such as pressure or temperature. The measurement of difference is the differential and is called the delta (as in delta P for pressure and delta T for temperature). The symbol is the Greek letter Δ.

Elasticity

The property of certain materials that enables them to return to their original dimensions after an applied stress has been removed. A typical example would be the difference in elasticity (modulus of elasticity) between steel and aluminum connecting rods. Aluminum stretches more than steel so you have to run more piston-to-cylinder-head clearance (than steel) to prevent component contact when the rod material stretches at TDC.

Equation

A mathematical statement asserting that two expressions are equal. It allows you to solve for unknown values by manipulating constants, known values, and variables according to mathematical laws to keep the equation balanced.

Exponent

A number or symbol placed to the right of and above another number, symbol, or expression to indicate the power to which it is raised. The number being raised is called the base number and the exponent indicates the number of times the base number is multiplied by itself. Hence 2^2 means 2 x 2 = 4 and 3^3 means 3 x 3 x 3 = 27. In engine math, numbers are typically squared (base number times itself once) or cubed (base number times itself twice).

Force

A vector quantity that tends to produce an acceleration of a body in the direction of its application. Example: In a cylinder, the combustion pressure (force) is constrained in all directions except one and that is the downward motion of the piston when combustion pressure (force) is applied.

Friction

The force that resists the motion of one surface relative to another with which it is in contact.

Fulcrum

The point of support upon which a lever pivots.

Inertia

The unique property of matter that causes it to resist any change in its motion or lack of motion (state of rest). Thus a body at rest remains at rest unless acted upon by an external force and a body in motion continues to move in a linear direction at constant velocity unless acted upon by an external force. Newton's first law of motion states that the mass of a body is a measure of its inertia.

Lever

The simplest mechanical device, consisting of a rigid bar pivoted about a fixed point and used to transmit a force. Leverage is the action or mechanical advantage of a lever and is represented by the amount of force applied over a given distance from the fulcrum to a load at a given distance on the opposite side of the fulcrum (or somewhere between the fulcrum and the point where force is applied).

Linear

Having only one dimension or direction.

Motion

A change in the position of a body or component with respect to time as measured by a particular observer in a particular frame of reference.

Percentage

A portion or share in relation to the whole where the whole is represented by 100. It is represented as a fraction or a ratio with 100 as the denominator. Example: 70/100 = 0.70 or 70 percent.

Pressure

The force acting normally on a unit area of surface, or the ratio of force to area. Example: Atmospheric pressure is 14.7 pounds per square inch (14.7 lbs/in^2) Absolute pressure (psia) is pressure measured on a gauge that reads zero at zero pressure rather than atmospheric pressure. Gauge pressure (psig) is measured on a gauge that reads zero at atmospheric pressure.

Ratio

A mathematical relationship in degrees or number between two things or quantities. In mathematics it is represented as a fraction. Example: In engine math the simplest ratio is that of crankshaft speed to camshaft speed. If the crankshaft turns at twice the speed of the camshaft, the ratio is 2:1.

Roots

Square roots ($\sqrt{x}$) and cube roots ($\sqrt[3]{x}$) are not often required, but it is good to understand that they are the reverse of exponents. They are tedious to calculate by hand. Fortunately you can find a square root or a cube root quickly on a scientific calculator by pressing the appropriate keys. If a number is squared 3^2 for example, the square root is expressed as $\sqrt{9}$ or 3. If a number is cubed (4^3), the cube root is expressed as $\sqrt[3]{64}$, or 4.

Specific Gravity

The ratio of the mass of a given solid or liquid to the mass of an equal volume of distilled water at 4 degrees C (39 degrees F) or of a gas to an equal volume

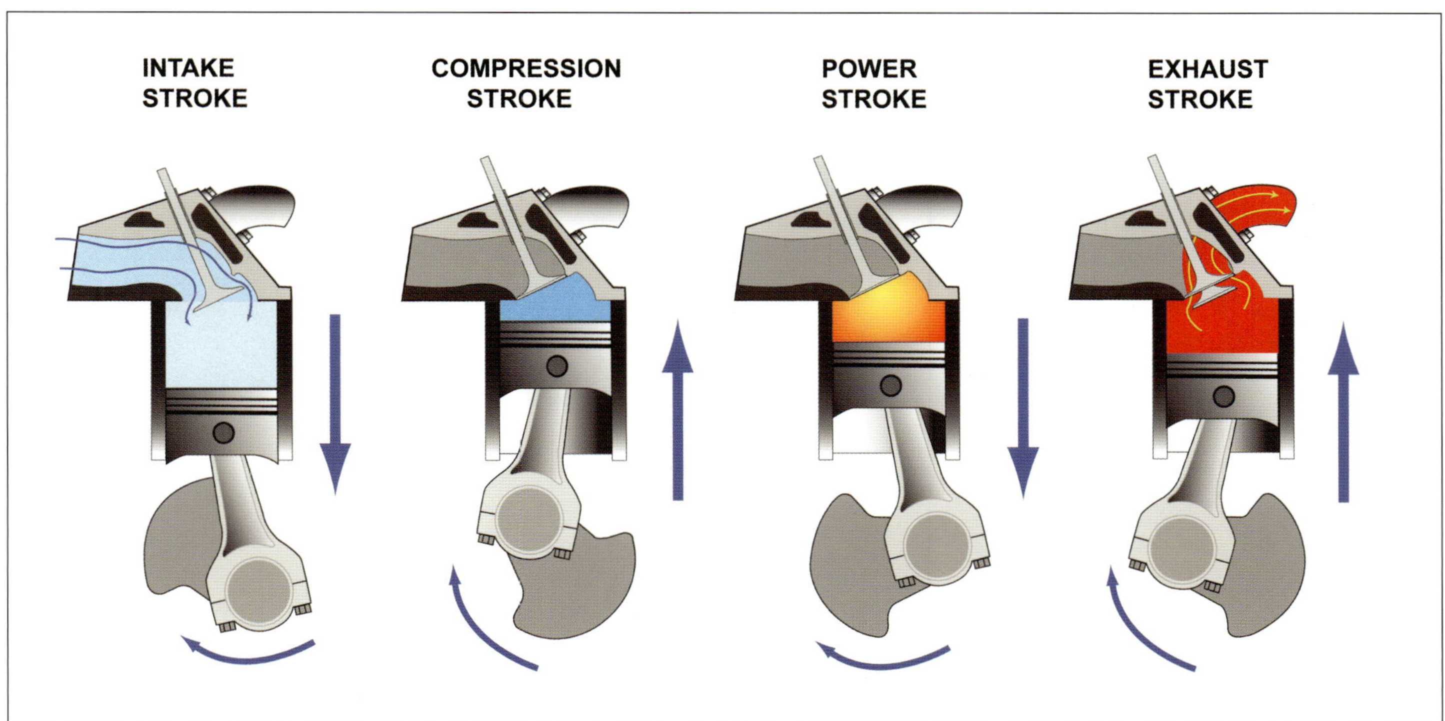

The four cycles of operation illustrate the relatively low pressures on the inlet and compression strokes versus the high pressures occurring on the power and exhaust strokes. Mathematical equations are used to describe and calculate the dynamic relationships that occur within the cylinders during each of these cycles.

of air or hydrogen under prescribed conditions of pressure and temperature.

Stoichiometric Ratio

The quantitive relationship between reactants and products in a chemical reaction. More specifically, in engine math the stoichiometric ratio of 14.7:1 represents the ideal air-fuel ratio for combustion (for gasoline) that creates minimal by-products.

Vector

A quantity such as velocity completely specified by magnitude (strength) and direction.

Volume

The amount of space occupied by a three-dimensional object or a region of space.

Variable

A quantity capable of assuming any assigned value or a symbol representing such a quantity.

If you had no problem grasping these basic concepts, you're ready to tackle any of the engine math calculations found in this book. Keep in mind that your calculations are only as accurate as your measurements. They say that numbers never lie and that is almost true. If you provide inaccurate measurements, the numbers won't lie, but they won't give the right answer either. Now grab your calculator and dive into the easy engine math formulas. They're going to be more useful and more fun than you expect.

Core Formulas

Area of a circle	$A = \text{diameter}^2 \times 0.7854$
Circumference of a circle	$C = \pi \times \text{diameter}$ or $2\pi r$
Volume of a cylinder	$V = \text{bore}^2 \times \text{stroke} \times 0.7854$
Area of a square	$A = \text{length} \times \text{width}$
Volume of a cube	$V = \text{length} \times \text{width} \times \text{height}$
Area of a right triangle	$A = 1/2 \times \text{base} \times \text{height}$

ENGINE DISPLACEMENT

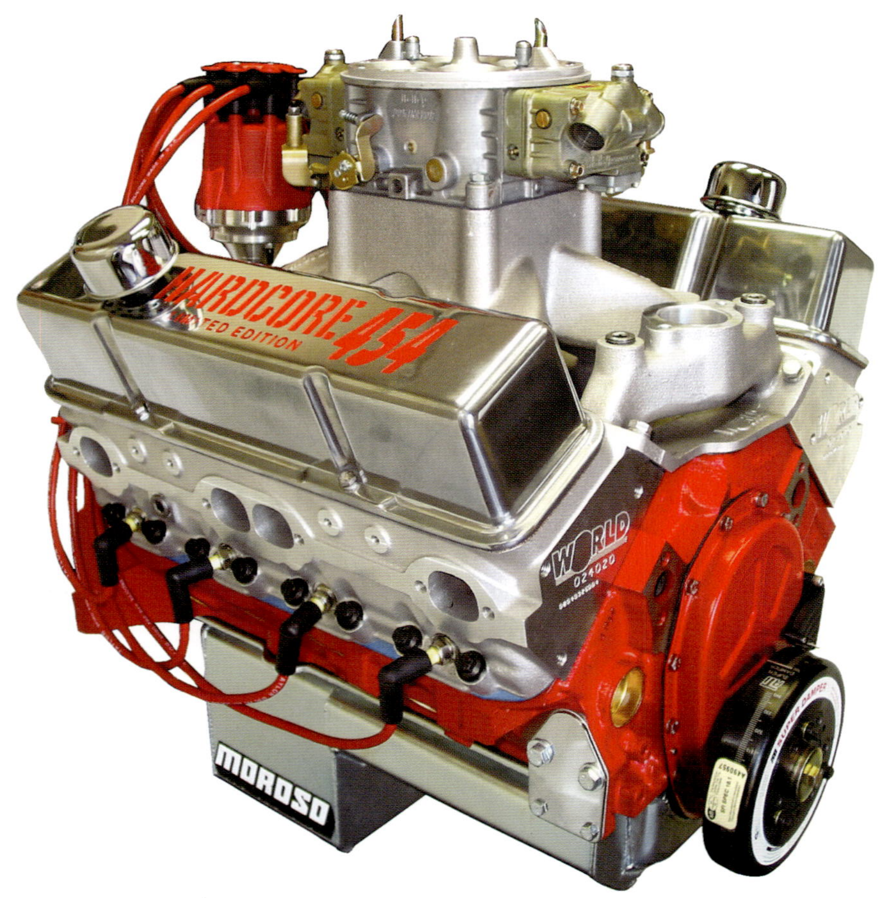

How does this World Products Hardcore 454 Chevy small-block cram all that cylinder volume into the same basic block architecture as a 283 Chevy? The bore is 4.250 inches and the stroke is 4.00 inches. Do the math.

Engine displacement is the most common math calculation. Displacement is the size or volumetric capacity of an engine expressed in cubic inches, cubic centimeters, or liters. Here in America we typically work in cubic inches while the rest of the world uses the metric system. I discuss appropriate conversions later in this chapter. Displacement is determined by a calculation involving the bore diameter and the stroke length times the number of cylinders. The re-sult is the actual swept volume of each cylinder and the total swept volume of the engine assuming 100 percent volumetric efficiency. Please note here that the actual swept volume is not the total volume of each cylinder since it does not include the volume of the combustion space above the piston at top dead center (TDC). These separate volumes allow you to calculate the engine's compression ratio (described in Chapter 3.)

Cylinder Bore Diameter

Cylinder bore diameter is a primary component of the engine displacement formula. Without a convenient comparison, any cylinder bore seems substantial to the eye, but even small changes in diameter relative to a fixed stroke length will produce an increase in engine displacement. Bore size is a major concern for any competition engine build because it dictates valve size and ultimately the breathing capability of the engine. Many engine builders feel that the breathing gains from a larger bore outweigh any friction penalties that may accrue from larger pistons with more skirt surface and potentially increased ring drag. A bigger bore also provides more piston area for combustion pressure to work against, but it also creates a greater distance for the flame front to travel and more surface area to cool the flame.

Street engines are one thing, but some racing series actually limit the bore size and bore spacing. These are typically cost measures designed to curtail the use of more expensive cylinder blocks with revised bore spacing, allowing larger bores while retaining desirable cylinder wall thickness and stability. Sprint Cup engines are a good example. The displacement is limited to 358 cubic inches with a maximum bore of 4.185 inches. Cup engines previously operated with a bore spacing of 4.400 inches, but NASCAR allowed a bore spacing increase to 4.5 inches to accommodate larger bores, bigger valves, and revised valve geometry—all in attempt to level the playing field among various brand competitors. If the cylinder bore is not specified, you must choose a bore dimension that best suits your particular application as defined by air flow and combustion chamber requirements, compression ratio, flame travel, and other factors, including a stroke length that also accommodates your operational requirements.

Precise bore measurements are taken with a dial bore gauge. Dial bore gauges typically read to an accuracy of ± 0.0005 inch and many are accurate to ± 0.0002 inch. It is no slight to any machinist for a customer to check their work, but it is important to recognize that your instruments may read different. That may be okay as long as your measurements fall within the acceptable tolerance. The best way to ensure the accuracy of any bore measurement is to set the tool according to a known standard prior to taking

Cylinder bores are measured with a dial bore gauge to obtain maximum accuracy. Measurements are taken at the top, center, and bottom of piston travel and in two different directions, front to back and side to side.

your bore measurements. If you have a favored machinist, you might also want to take some of your instruments by his shop and compare sample measurements with his tools.

Stroke Length

Stroke length is the companion factor in the cylinder displacement formula. Adding stroke length increases displacement relative to a fixed bore size. Stroking, an early hot rodding trick, has found particular favor in many late-model engine builds seeking to maximize displacement. With the exception of high-performance applications, stroke lengths usually remain fixed with the factory length as it is much easier and more practical to increase the

<table>
<tr><td>

Critical Short Block Dimensions

Cylinder Block
- Block Deck Height
- Bore Spacing
- Bore Diameter
- Main Bearing Housing
- Bore Diameter

Crankshaft
- Stroke
- Main Journal Diameter
- Rod Journal Diameter

Connecting Rod
- Center-to-Center Length
- Pin Diameter
- Big End Diameter

Piston
- Piston Diameter
- Piston Pin Height
- Pin Diameter

</td></tr>
</table>

Many racing applications find it beneficial to use the largest bore possible in order to gain more piston surface area and unshroud the valves to promote more efficient breathing. For example, a 4.125–inch bore typically promotes better breathing than a 4.00–inch bore because it provides a more efficient flow path for the intake charge. If you are displacement limited, you can use a variant of the displacement formula to calculate the appropriate stroke required by your bore selection or vice versa. More about this later.

Calculating Displacement

To find total displacement, begin by calculating the volume of a single cylinder and then multiply the result by the number of cylinders in the engine. The formula for the volume of a cylinder requires the use of pi, a mathematical constant (see sidebar "How the Displacement Formula Works" on page 15) that allows the calculation of the area (or volumetric difference) between a square (or a box) and a circle (or a cylinder). Pi divided by 4 gives you another constant to complete the basic displacement formula as follows:

displacement via bore enlargement. And in the case of your typical engine rebuild, the primary concern is to restore cylinder sealing with new oversized pistons and rings. A simple bore-and-hone job is all that's required. Stroke increases often require block modifications to provide clearance for the rods and rod bolts, and require the purchase of new pistons with the appropriate pin location to accommodate the new stroke length. In either case, the displacement formula can be manipulated to calculate displacement or to find the required bore or stroke when the desired displacement and one of the dimensions is known.

To calculate an engine's displacement you must first find the swept volume of an individual cylinder based on the bore and stroke dimensions. The bore is the diameter of the cylinder and the stroke is the distance that the piston travels up and down in the cylinder. (Stroke is actually a function of the length of the crankshaft throw, but it is commonly referenced by the travel of the piston top.)

In almost every case you will already know the size of the engine and the bore and stroke dimensions, but calculating engine displacement with precision allows you to determine the amount of "rounding" the factory has applied to the stated displacement. Sometimes they round up; sometimes they round down. In practice, a more useful reason is the ability to brainstorm various engine configurations to suit a particular racing class that specifies a displacement limit, or to calculate the displacement effect of over-boring on an engine rebuild.

The 572-ci Merlin big-block uses a unique combination of bore and stroke to achieve its massive displacement. In this case we're looking at a square engine with a 4.500-inch bore, a 4.500-inch stroke and 735 hp. Another version adds 1/4 inch more stroke to achieve 632 ci and 800 hp on pump gas. (Courtesy World Products)

How the Displacement Formula Works

Like all the formulas in this book, the displacement formula incorporates variables. Variables are plug-in values or actual dimensions that you can measure and apply to any formula. The formulas are all algebraic equations.

We know the formula for displacement is:

$$\text{bore}^2 \times \text{stroke} \times 0.7854 \times \text{number of cylinders}$$

variable   *variable*   *constant*   *variable*

Algebraically this would be: $a^2 \times b \times 0.7854 \times c$

Where:
a = bore size
b = stroke length
c = number of cylinders
0.7854 = a constant

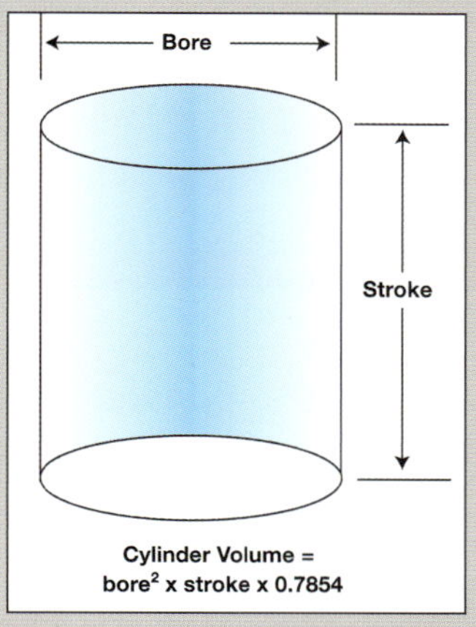

This is the same formula for calculating the volume of a box or a cube, with the exception of the pi-derived constant, which converts the calculation to a cylinder instead of a cube. Let's work an example for each, assuming that the engine's cylinders are actually square in the second case. For these examples I'll use a 4-inch bore and a 3-inch stroke for a 302-ci Chevy.

Cylinder Displacement =
bore x bore x stroke x 0.7854 x number of cylinders

Cylinder Displacement =
4 x 4 x 3 x 0.7854 x 8 = 301.59, or 302 ci

Cube Displacement =
length x width x height x number of cylinders

Cube Displacement = 4 x 4 x 3 x 8 = 384 ci

A square box or cube of any equivalent height has more volume than a cylinder because it has square corners. Pi gives us the constant (0.7854) that converts the difference. If we now divide 301.59 by 384 we get 0.7853906 (or 0.7854 when rounded). Move the decimal point two places to the right and it is easy to visualize that a cylinder's volume is

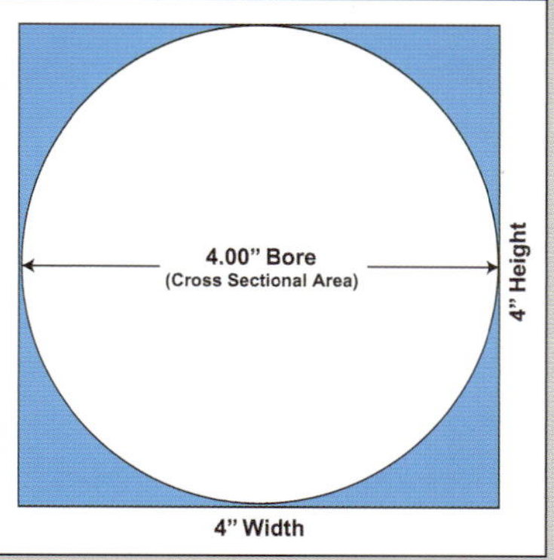

always 78.54 percent of the square cube volume regardless of the height or stroke dimension. Neat huh?

To find the area of the square, multiply length x width.
4 x 4 = 16 sq. inches

To find the area of the circle use pi/4 or 0.7853982.

4 x 4 x 0.7853982 = 12.5663712 sq. inches

Divide 12.5663712 by 16 and you get 0.7853982.

In practice, pi/4 is rounded to 0.7854

So in our example:

4 x 4 x 0.7854 = 12.5664 or 12.5664 ÷ 16 = 0.7854

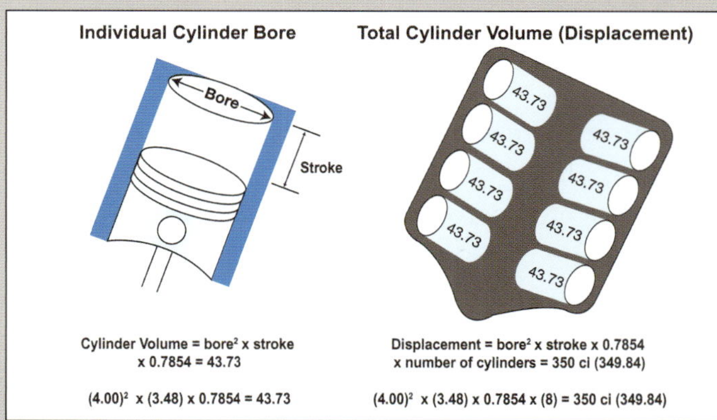

The swept volume of a single cylinder is determined by the bore and stroke dimensions. To get total engine displacement, multiply the volume of a single cylinder by the number of cylinders.

Converting Fractions

Converting fractions to decimal notation is simple. A fraction is just what it looks like: one number divided by another. The top number is the numerator and the bottom number is the denominator. When they are the same the number is equal to 1. In decimal notation that is 1.000 with the decimal point carried out to three places (or more in some cases). Measurements are usually taken in thousandths of an inch, which is three places to the right of the decimal point. A typical measurement might be 4.060 inches. You might recognize that as an overbore of 0.060, or sixty thousandths.

Since engine math frequently works in decimals, you often have to convert fractions to decimal notation. The conversion is simple. Just do the division as indicated by the fraction. The top number is divided by the bottom number.

$$16 \div 16 = 1 \text{ or } 1.000$$

$$27 \div 32 = 0.843$$

If you have a scientific calculator with parenthesis keys you can incorporate the fractions into your calculations by substituting the plus sign (+) between the integers and the fractions. Here's a sample displacement calculation using a 4⅛-inch bore and a 3⅝-inch stroke. Press the keys exactly as shown here:

$$.7854 \text{ x } (4⅛)^2 \text{ x } (3⅝) \text{ x } 8 = 387.55 \text{ ci}$$

Note that when dividing you actually press the ÷ key. To square the bore dimension, press the x^2 key.

$$Pi\ (\pi) = 3.1415927$$

$$\text{Cylinder Volume} = pi \div 4 \text{ x bore}^2 \text{ x stroke}$$

$$Pi \div 4 = 0.7853982 \text{ (commonly rounded to 0.7854)}$$

Hence, 0.7854 becomes the constant that helps us calculate cylinder volume.

The formula for displacement is:

$$\text{Bore}^2 \text{ x stroke 0.7854 x number of cylinders}$$

Let's work an example: If a 327 Chevy engine has a published bore of 4.00 inches and a 3.25-inch stroke, the calculation becomes:

$$4.00^2 \text{ x } 3.25 \text{ x } 0.7854 \text{ x } 8 = 326.726 \text{ ci}$$

While infrequent, sometimes you will find that an engine's dimensions are listed with a common fraction. In the case of our Chevy example, it is often referred to as having a three-and-a-quarter-inch (3¼) arm or crank. Since engine math works in decimal notation,

you need to convert that fraction to a decimal. The following chart shows the most common examples.

Fraction	Decimal
1/32	0.031
1/16	0.0625
1/8	0.125
1/4	0.250
3/8	0.375
1/2	0.500
5/8	0.625
3/4	0.750
7/8	0.875
1	1.000

For our 327 Chevy with a 3¼-inch crank we note that 1/4 inch equals 0.25 inch, so the stroke in decimal notation is 3.25 inch; hence the following calculation:

$$4.00^2 \text{ x } 3.25 \text{ x } 0.7854 \text{ x } 8 = 326.726 \text{ ci}$$

In this case Chevrolet rounded up to 327 ci because the decimal is more than 326.5, or closer to 327 than 326. Manufacturers don't always follow this practice. They sometimes round in the opposite direction even if the decimal doesn't require it. This is frequently done to differentiate among engine families or newer models.

Calculating Overbore Displacement

Suppose we decide to overhaul our 327 Chevy. An examination reveals excessive cylinder wear and a noticeable ridge at the top of each bore above the area of piston ring travel. We decide to install new pistons that are 0.030-inch oversize. What will the displacement be on our newly rebuilt 327-ci engine? This is easy to calculate by plugging the new bore value (a variable) into the standard displacement formula:

$$4.030^2 \times 3.25 \times 0.7854 \times 8 = 331.645 \text{ ci}$$

While this would normally round up to 332 ci, for some unknown reason this particular overbore is commonly called a 331 on a Chevy small-block. Strangely, substituting for a 0.060-inch overbore yields 336.601 ci, which is commonly called a 337. Go figure.

Calculating Bore and Stroke Relationships

The next two formulas are most useful for brainstorming engine combinations that have a displacement limit. Why is this important? Most often, it allows you to calculate a stroke length to suit a desirable bore size for a fixed displacement set by a racing sanctioning body. This is critical to keeping your engine size within the maximum allowable displacement.

The NHRA for example, always rounds up to the next largest number, so anything greater than the stated engine size would make your engine illegal. At Bonneville, the SCTA publishes an engine displacement range, which savvy racers push to the limit in each class. A C-class engine, for example, can range from 306.00 to 372.99 ci. They won't round up to the next inch, but your calculation had better not exceed 372.99 by any margin. Most people shoot for about 1 ci less, just to provide a safety margin. So if you are building a C-class Bonneville engine with a small-block Chevy and you feel that a 4.125-inch bore will breathe better than a 4.00-inch bore, what stroke length will you need to meet the 372.99-ci limit?

This all-aluminum LS7-based Warhawk small-block takes maximum advantage of the Gen III engine's 0.780-inch taller deck height of 9.800 inches to stuff a 4.5-inch stroke into the 4.125-inch bore block to achieve 481ci within the late-model small-block architecture. (Courtesy World Products)

Stroke Length when Bore is Known

If you know the bore size and the final displacement you can use them to calculate the stroke length. This is useful for brainstorming bore and stroke combinations where you wish to examine various possibilities for achieving a desired displacement.

$$\text{Stroke} = \text{displacement} \div$$
$$(\text{bore}^2 \times 0.7854 \times \text{number of cylinders})$$

$$\text{Stroke} = 372.99 \div (4.125^2 \times 0.7854 \times 8)$$
$$= 3.4887 \text{ inches}$$

What this tells you is that you could use a standard 400-ci Chevy bore dimension (4.125-inch) with a standard 350 Chevy (3.48-inch stroke) crankshaft to stay under the limit. A 350 Chevy stroke is 3.48 inches, which computes to 372.055 ci; a little close for comfort, but possible if you maintain your dimensions carefully. This would keep your crankshaft expense in check by using an off-the-shelf crank size.

What is Pi?

Pi (π) is the ratio of the circumference of a circle to its diameter. It is a mathematical constant that remains the same regardless of the size of the circle.

Engineers typically use pi carried out to five significant digits (3.1416) for accuracy and it has been shown that a value taken to eleven decimal places is sufficient to calculate the circumference of the earth to within 1 mm. Three digits (3.14) suffice for most requirements.

$$\pi = \text{Circumference} \div \text{diameter, or } C \div d$$

This is often referred to as the circular constant.

Pi can also be defined as the ratio of a circle's area (A) to the area of a square whose side is equal to the radius of the circle. Note that the sector scribed by the radius of the circle is 1/4 of pi. If you think about it, that's how we get the displacement constant 0.7854, which is pi divided by 4. In effect, the length of that sector represents one side on an imaginary square smaller than a square with sides the length of the circle diameter. That represents the area of the circle compared to a square of the same length and width as the circular diameter.

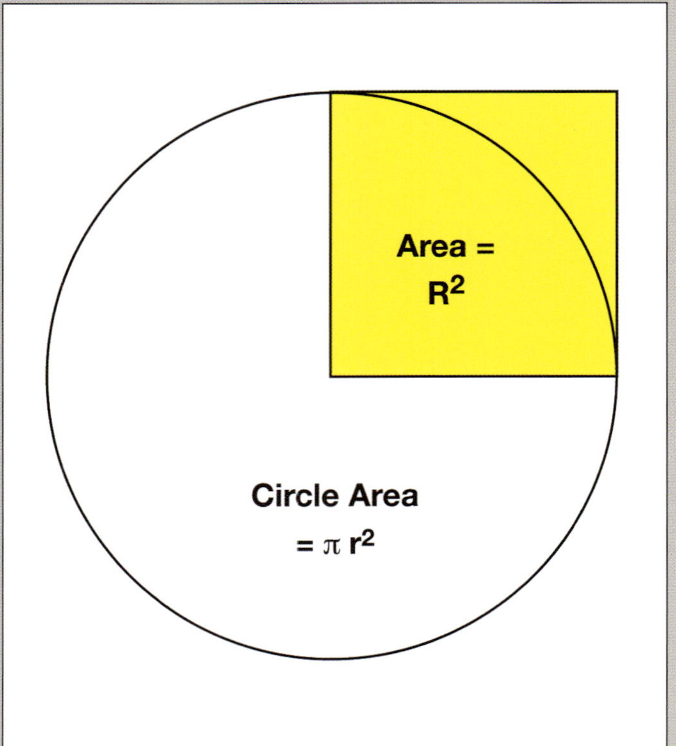

The shaded area inside the square represents the difference in area between the square and the circle (bore). The same applies to volume if you add a height dimension (stroke).

Let's say you want to further aid the engine's breathing and piston area by increasing bore size to 4.155 inches.

$$\text{Stroke} = 372.99 \div (4.155^2 \times 0.7854 \times 8)$$
$$= 3.4385 \text{ inches}$$

That's an odd crank size that requires special grinding. Sure, a crank grinder could make it, but you have to weigh the potential airflow benefit versus the crank grinding cost plus the expense of purchasing pistons with a pin bore height that matches the odd stroke and your desired rod length. In some cases, engine builders use this same calculation to see how much they can shorten the stroke to gain more RPM potential with less piston speed. In either case, the formula for finding stroke allows you to brainstorm ideas on paper before you lay down cold cash for parts.

Bore when Stroke is Known

Some times you have a particular stroke in mind and you want to calculate a bore dimension to suit a given displacement. I recently had occasion to brainstorm some small-displacement Chevy V-8 combinations that had a cubic inch limit of 260.99. My first thought was to use a small-displacement block from a 283 or a 305. I knew the stroke lengths for both of them, but I couldn't recall the bore diameter of the 305 and I didn't have a handy reference. Fortunately, I was able to calculate it by using the formula for finding an unknown bore dimension when the stroke is known.

To work the formula, do the math inside the square root symbol first, then enter the result into your calculator and press the square root key to get the answer.

$$\text{Bore} = \sqrt{[\text{displacement} \div (0.7854 \times \text{stroke} \times \text{number of cylinders})]}$$

In the case of a 305 Chevy, the stroke is 3.48 inches, the same as a 350. By plugging in the known variables, you calculate the unknown bore dimension as follows.

$$\text{Bore} = \sqrt{[305 \div (0.7854 \times 3.48 \times 8)]}$$
$$= 3.734 \text{ inches bore diameter}$$

Calculator sequence:

$$305 \div (0.7854 \times 3.48 \times 8) = \sqrt{13.948} = 3.734$$

I was then able to use that dimension to calculate a stroke that would yield 260.99 ci. To do this, I went back to the formula for finding stroke.

$$\text{Stroke} = \text{displacement} \div$$
$$(0.7854 \times \text{bore}^2 \times \text{number of cylinders})$$

$$\text{Stroke} = 260.99 \div (0.7854 \times 3.734^2 \times 8)$$
$$= 2.979 \text{ inches}$$

Neither dimension is convenient. The bore is too small to breathe well and the stroke is an odd number that will have to be specially ground. After further consideration, the logical choice is a 4.00-inch-bore block, which is readily available. It offers the best breathing, and with the stated displacement limit, there is no way around having to get a crank ground to the required stroke dimension. Using the 4.00-inch-bore dimension the stroke requirement computes as follows.

$$\text{Stroke} = 260.99 \div (0.7854 \times 4.00^2 \times 8)$$
$$= 2.596 \text{ inches}$$

The interesting thing here is the temptation to round off the stroke to 2.600 inches. Let's see what happens when we plug that into the displacement formula.

$$\text{Displacement} = 4.00^2 \times 2.600 \times 0.7854 \times 8 = 261.38 \text{ ci}$$

A negative deck measurement where the piston top is above the block deck surface at TDC does not alter displacement because the calculation is still based on the bore and stroke dimensions. It affects your compression ratio calculations (as discussed in Chapter 3).

Uh-oh! Busted for an oversize engine.

Let's try the exact stroke figure we calculated previously.

$$\text{Displacement} = 4.00^2 \times 2.596 \times 0.7854 \times 8$$
$$= 260.978 \text{ ci}$$

You can't get much closer to the displacement limit than that, but it's a little too close. A more comfortable choice would be to shorten the stroke to 2.585, which yields roughly a 1-ci safety margin.

$$\text{Displacement} = 4.00^2 \times 2.585 \times 0.7854 \times 8 = 259.87 \text{ ci}$$

You can see from this exercise how you can use the displacement formula, bore formula, and stroke formula to model theoretical combinations that meet your requirements. Once you settle on a bore and stroke combination that meets your displacement limit with the best possible breathing, piston area, and RPM capabilities, you can move on to selecting the best connecting rod length and piston pin height to match. These handy formulas make it all happen on your calculator and on paper before you spend a dime on parts. And, if you enter the variables into a computer spreadsheet (discussed in Chapter 13), you can save and print all your theoretical combinations for quick review whenever you wish.

Bore/Stroke and Rod/Stroke Ratios

Another thing to consider during your brainstorming sessions is the bore/stroke ratio and the rod/stroke ratio. These ratios tell you several important things. If an engine has the same dimensions for bore and stroke it is said to be square. If the bore is greater than the stroke, the engine is over-square. An under-square engine would have a stroke length greater than its bore dimension. Low-speed under-square engines are often thought to promote more torque, but in practice, the difference is minimal and can't compete with the superior breathing of an over-square combination. While a larger bore may have slightly higher friction properties, it more than overcomes the difference with the improved breathing offered by a larger bore size that permits larger valves and more efficient air/fuel charge entry into the cylinders. The bore/stroke ratio is simply the bore dimension divided by the stroke.

$$B/S \text{ ratio} = \text{bore} \div \text{stroke}$$

For example, a Chevy 350 engine has a standard bore/stroke ratio of 1.15:1,

$$B/S = 4.00 \div 3.48 = 1.15{:}1$$

Any number above 1.0:1 is over-square. A larger bore permits larger valves. Consider the following examples based on three different Chevy small-blocks:

Displacement	283 ci	350 ci	400 ci
Bore	3.875	4.000	4.125
Stroke	3.000	3.480	3.750
B/S Ratio	1.29:1	1.15:1	1.1:1

With the highest ratio, the 283 looks okay on paper. It is efficient for its displacement and stroke length, but the small bore restricts valve size. The 350 has a lower B/S ratio, but its bore permits much larger valves than the 283 so breathing is more efficient. The 400 is almost square at 1.1:1, but it is the most desirable from a breathing standpoint because it can accept monster valves without shrouding or restricted breathing. It's one more thing to consider when you are planning your optimum combination.

Another relationship to consider is the rod/stroke ratio. This is the rod center-to-center length divided by the stroke. A rod/stroke ratio of 1.9 to 2:1 has always been considered beneficial because it reduces the angle of the rod to the cylinder wall, thereby lowering the side-loading on the cylinder wall and the piston skirt. That's why the short rod (5.565 inches), long stroke (3.75 inches) Chevy 400 small-blocks experienced higher cylinder wear. The rod angularity was too great. A 302-ci Chevy (with its 5.7-inch rod and a 3.00-inch stroke) has a rod/stroke ratio of 1.9:1. Compare that to the Chevy 350 and 400 small blocks shown in the following chart.

$$R/S \text{ Ratio} = \text{rod length} \div \text{stroke}$$

Displacement	302 ci	350 ci	400 ci
Rod Length	5.7	5.7	5.565
Stroke	3.00	3.48	3.75
R/S Ratio	1.9:1	1.64:1	1.48:1

A longer rod and higher rod/stroke ratio is also purported to improve power by "parking" the piston at TDC for a fraction longer than a short rod ratio. This allows more time for peak cylinder pressure to build before the piston starts downward on the expansion cycle. Comparative dyno tests suggest this benefit to be less substantial than theory promises, at least at street engine speeds. However, the durability benefits of improved rod angularity are more apparent. In any case, you may want to compute the bore/stroke ratio and the rod/stroke ratio of any combination you design and consider the implications.

Metric Conversions

If you are dealing with metric dimensions, you are looking at bore and stroke dimensions measured in millimeters (mm) and engine displacement stated in cubic centimeters (cc) or liters (L). The formula for displacement works the same way, but you divide the result by 1,000 to obtain cubic centimeters.

Sample Displacement Calculations

These practice calculations are based on published bore and stroke dimensions for three current model high performance musclecars. Try working the formula in inches to get cubic inches and then in millimeters to get cubic centimeters and liters.

2010 Camaro SS	**Bore x Stroke**
L99 V-8, 376 ci, 6.2L	4.06 x 3.622 inches
426 bhp @ 5,900 rpm	103.3 x 92 mm
420 ft-lbs @ 4,600 rpm	Over-square engine

$$4.06^2 \times 3.622 \times 0.7854 \times 8 = 374.92 \text{ ci}$$
$$10.33^2 \times 9.2 \times 0.7854 \times 8 = 6168.35 \text{ cc}$$

The L99 stroke is actually 3.622 inches, which computes to 375.12 ci. Chevy rounds this up to 376 ci. To get liters, divide 6,168 by 1,000. The result is 6.168L which Chevy rounds up to 6.2L. Note that 376 ci divided by the conversion factor (61.024) yields 6.161L; almost identical to the metric calculation.

2010 Dodge Challenger	**Bore x Stroke**
Hemi V-8, 345-ci, 5.7L	3.92 x 3.58 inches
376 bhp @ 5,200 rpm	99.5 x 90.9 mm
410 ft-lbs @ 4,300 rpm	Over-square engine

$$3.92^2 \times 3.58 \times 0.7854 \times 8 = 345.64 \text{ ci}$$
$$9.95^2 \times 9.09 \times 0.7854 \times 8 = 5654.45 \text{ cc}$$

Note that Dodge rounds down to 345 ci, but the cubic centimeters are rounded up to get a 5.7L engine. Divide 5,654.45 by 1,000 and get 5.654L, appropriately rounded to 5.7L. Now divide 345 ci by the conversion factor (61.024) and you get 5.653L. Again, it's very close to the metric calculation and appropriately rounded to 5.7L

2010 Mustang GT	**Bore x Stroke**
Modular V-8, 281 ci, 4.6L	3.55 x 3.54 inches
315 bhp @ 6,000 rpm	90.2 x 90 mm
325 ft-lbs @ 4,250 rpm	Nearly square engine

$$3.55^2 \times 3.54 \times 0.7854 \times 8 = 280.31 \text{ ci}$$
$$9.02^2 \times 9.0 \times 0.7854 \times 8 = 4,600.83 \text{ cc}$$

Note that 4,600.83 cc divided by 1,000 equals 4.6L. Also note that Ford rounds 280.31 ci to 281 ci. Divide 281 ci by the conversion factor (61.024) and you get 4.593L. Again, close enough to be called a 4.6L engine.

For example:

5,000 cc = 5.0L
a 305 Chevy = 305 ÷ 61.024 = 4,998 cc = 5.0L
a 302 Ford = 302 ÷ 61.024 = 4,948 cc = 5.0L

Note the subtle difference.

In the Introduction, I suggested that repetition is one of the best ways to learn something. After using your calculator to practice the calculations and conversions shown above, you will have a pretty good feel for how the numbers are manipulated. One important thing to remember if you are racing is that measured dimensions are always true. Published dimensions are subject to rounding and conversion skew, so judge accordingly.

$$\text{Displacement}_{\text{liters}} =$$
$$(\text{bore}^2 \text{ [mm]} \times \text{stroke [mm]} \times 0.7854 \times 8) \div 1,000$$

No conversion is required for the constant (0.7854) because it applies regardless of the unit of measure. To gain a better understanding of it, let's work the displacement formula for a hot contemporary engine like the 426-hp L99 aluminum small-block in the 2010 Camaro. It has the following dimensions:

Stated Dimensions	**Calculated Dimensions**
L99 Chevy 376 ci, 6.2L	375.129 ci, 6.147L
Bore: 4.06/103.3 mm	103.12 mm
Stroke: 3.622/92 mm	91.998 mm

To convert inches to millimeters, multiply by 25.4 (see Appendix B "Handy Conversion Factors"). This conversion gives an exact number so it is very accurate. To convert from millimeters to inches, multiply by 0.0393701, which does not yield an exact number, but it is close enough.

Let's work with inches first.

$$\text{Displacement} = 4.06^2 \times 3.622 \times 0.7854 \times 8 = 375.129 \text{ ci}$$

In this case Chevy rounded up to 376 ci. Now let's convert to millimeters.

$$\text{Bore} = 4.06 \times 25.4 = 103.124 \text{ mm}$$
$$\text{Stroke} = 3.622 \times 25.4 = 91.998 \text{ mm}$$

Since converting from inches to millimeters is exact, the figures are dead accurate, but here we'll round them off because they are rounded in most published accounts. Most often the bore is stated as 103.3 mm and the stroke is quoted at 92 mm.

To simplify our calculations let's round the numbers by reducing 103.124 mm to 103 mm and increasing the stroke to 92 mm.

$$\text{Displacement}_{cc} =$$
$$(103^2 \times 92 \times 0.7854 \times 8) \div 1,000 = 6,132.57 \text{ cc}$$

Divide 6132.57 cc by 1,000 to get liters—6.132. So, mathematically the L99 is 6,132 cc or 6.13L (which Chevy rounds up to 6.2L). Note that the cubic centimeters and liters converted from inches are slightly different from those calculated from millimeters.

If your measurements are already in millimeters, you can simplify by converting to centimeters before you work the calculation. This will yield an answer in centimeters without having to divide by 1,000. To convert the bore and stroke dimensions to centimeters, divide each number by 10.

$$\text{Bore} = 103 \div 10 = 10.3 \text{ cm}$$
$$\text{Stoke} = 92 \div 10 = 9.2 \text{ cm}$$

$$\text{Displacement} = 10.3^2 \times 9.2 \times 0.7854 \times 8 = 6,132.57 \text{ cc}$$

This is identical to our previous calculation.

If you are brainstorming combinations in metric units, the basic formulas for displacement, bore and stroke still apply. You just have to be careful to keep your units straight and work any conversions properly. (Consult the conversion chart in Appendix B).

You can see that published dimensions do not always match measured dimensions. A tech inspector always goes by the actual measured dimensions in the units called for by the rulebook. If you are dealing with a sanctioning body, always use the units they specify to ensure a legal engine build.

Another convenient conversion to remember is from cubic inches to liters. To convert, divide the cubic inches by 61.024. For our L99 Chevy, the calculated displacement was 375.12/61.024 equaling 6.147 liters (which Chevy rounds up to 6.2L). We already calculated the cubic centimeters at 6,132.57. Dividing by 1,000 yields 6.132L. That is still pretty close to exact, but you can see how rounding and conversion factors can skew the final answer.

Equivalent Displacement

There is a formula for calculating equivalent displacement for a non-reciprocating engine such as a Mazda rotary. This helps classify non-reciprocating motors against conventional piston engines.

$$\text{Equivalent Displacement} = SV \times 3$$

Where:
SV = the actual swept volume of one single chamber of the rotary engine.

In this case you would substitute the published swept volume since you can't bore and stroke a rotary engine.

What Is Bore Spacing?

Bore spacing is the distance between the centerlines of adjacent cylinder bores. It helps determines how much you can overbore a cylinder block without weakening the cylinder walls to the point of potential failure. As shown, a small-block Chevy has a bore spacing of 4.400 inches. Note the radius of the adjacent bores in the illustration and you see that there is 0.400-inch between the cylinder bores for

Calculator Tips

Here are some keypad tips for working calculations that have square root symbols and or numbers enclosed by parenthesis.

Let's use the formula for stroke as an example:

$$\text{Stroke} = \text{displacement} \div (\text{bore}^2 \times .7854 \times \text{number of cylinders})$$

First enter the displacement. Then press the divide key (÷). Press the open parenthesis key [(], enter the bore 0.00, then press the square key [x^2], then x for times, 0.7854, x for times, number of cylinders 8, closed parentheses [)], and then press the equal key [=].

Now, try again with the formula for bore:

$$\text{Bore} = \sqrt{[\text{displacement} \div (.7854 \times \text{stroke} \times \text{number of cylinders})]}$$

Enter displacement, press the divide key [÷], press the open parenthesis key [(], then 0.7854, x for times, enter stroke 0.00, x for times, number of cylinders [8], closed parentheses [)] , press the equal key [=], and then press the square root key [√].

Note that by pressing the parenthesis keys to enclose operations you can work through the formula just the way it appears.

cylinder walls and water jacketing. The cylinder walls are typically 0.180- to 0.200-inch thick with very little space for the cooling jacket between them.

The cylinder walls are actually egg shaped in the water jacket and they taper between the bores to accommodate coolant flow. This area is tangent to piston thrust and is not subjected to maximum loading. On large-bore blocks such as the 400 Chevy, adjacent cylinder walls are joined or siamesed with block material to maintain strength.

When you bore a block you are taking one half the over bore measurement out of each side of the cylinder. On a 0.030-inch overbore, you're only losing 0.015-inch material overall. Most blocks accept up to 0.060-inch overbore without loss of cylinder wall strength, except for GM's newer LS-series engines. These use an iron sleeve that can only be over-bored 0.010-inch. In practice it doesn't matter what kind of engine you have. Your machine shop will be very familiar with the safe boring limits of most engines and can even sonically check the cylinders to determine if they are thick enough to accommodate your desired bore dimension and the final application of the engine.

Some engines (like the 305-ci Chevy, for example) are built with thin-wall castings to reduce engine weight. In theory they will never be subjected to high RPM or racing loads, so thicker walls are unnecessary. These blocks should be sonic-checked prior to any substantial overbore.

In contrast, the old 283 Chevy blocks were known for thick cylinder walls that would accept an overbore of 0.125 inch. This yielded a 4.00-inch bore and the

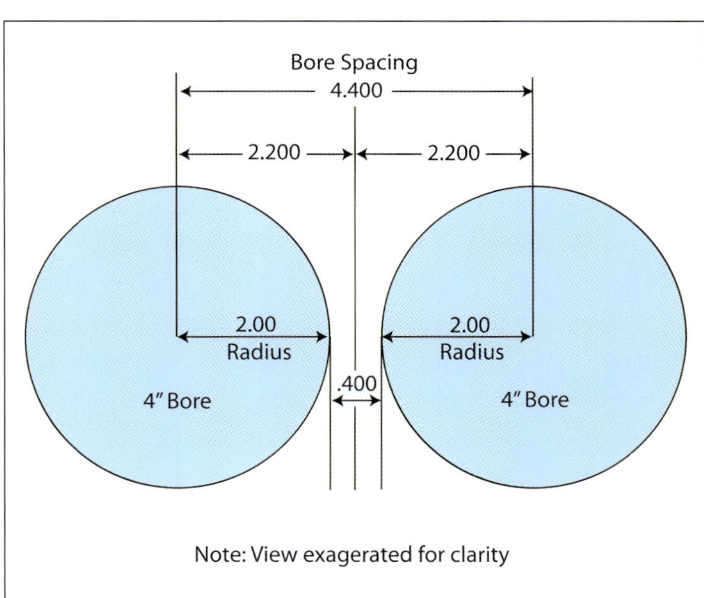

Bore spacing is the fixed distance between the centerlines of adjacent cylinder bores. Bore size may change, but bore spacing remains constant. As shown here, the bore spacing dimension can hep you determine how much you can safely over-bore your block.

Calculating Deck Height and Pin Height

While deck height is commonly referred to as the depth of the piston top in the bore when the piston is at top dead center, it also refers to the distance from the crankshaft centerline to the deck surface of the cylinder block and influences all combinations of stroke, rod length and pin height. A simple formula is used to calculate the combined dimensions for comparison with the cylinder block's fixed deck height dimension:

$$\text{Assembled Deck Height} =$$
$$\text{block height} - [(\text{stroke} \div 2) + \text{rod length} + \text{pin height}]$$

Where:
Block height = fixed deck height of the cylinder block
Stroke = crankshaft stroke length
Rod length = connecting rod center-to-center length
Pin height = pin centerline depth from piston top
(commonly called compression height)

Since block height is fixed within a narrow window available for deck milling, the combination of stroke length, rod length, and pin height must add up to the same height with a small tolerance for desired deck height and piston-to-head clearance, which also incorporates gasket thickness. A common practice in performance circles is to zero-deck the block. That means that the combination of one half the stroke length plus rod length and pin height equals the fixed deck height of the block. The flat portion of the piston top is exactly even with the deck surface of the block. This forces the builder to select the appropriate compressed gasket thickness to control piston-to-head clearance. Not surprisingly, most performance head gaskets are 0.039- to 0.042-inch thick when compressed. The commonly accepted minimum piston-to-head clearance with steel connecting rods is 0.035 inch; hence, the built-in fudge factor on the head gaskets which averages about 0.005 inch.

This calculation is often used to help determine the required piston pin height to accommodate a desired rod length and stroke combination, particularly when there is a concern that the selected rod length might push the pin location up into the ring package. To illustrate this we'll examine a typical 350-ci small-block Chevy V-8, which has a nominal factory block height of 9.020 inches. The factory stroke length is 3.48 inches, the stock connecting rod measures 5.7 inches, and the factory pin height is 1.560 inches. Plugging in our variables, we find the following:

$$\text{Assembled Deck Height} =$$
$$\text{block height} - (\text{stroke} \div 2 + \text{rod length} + \text{pin height})$$

Or
$$ADH = 9.020 - [(3.48 \div 2) + 5.7 + 1.56]$$
$$ADH = 9.020 - (1.74 + 5.7 + 1.56)$$
$$ADH = 9.020 - 9.00 = 0.020 \text{ inch}$$

Stock small-block Chevy pistons are indeed about 0.010 to 0.020 inch down the bore from the factory. This small amount of deck clearance allows for minor deck milling to recondition the deck surface for future rebuilds. Now suppose you want to build this into a performance engine with longer connecting rods and new pistons. Component sources are everywhere. We selected a 0.030-inch oversize flat-top piston from the Howard Cams catalog, PN HRC4567. This particular piston accommodates a 6-inch connecting rod by virtue of a 1.268 pin height. It is also designed to work with either a 3.48- or 3.50-inch pin. Let's see how that works out.

$$ADH = (\text{stroke} \div 2) + \text{rod length} + \text{pin height}$$
$$1.74 + 6.00 + 1.268 = 9.008 \text{ inches}$$

If your original block is indeed 9.020 inches from the crank centerline to the deck surface, you can cut the decks up to 0.012 inch to make them parallel and achieve a zero-deck assembly. To use this particular piston with a 3.50-inch stroke we find that the assembly adds up to 9.018 inches, only 0.002 inch less than the nominal block height of 9.02 inches.

1.75 + 6.00 + 1.268 = 9.018 inches

If your block requires more than 0.002-inch milling to make the decks parallel, you may end up with some degree of negative deck (piston protruding from the bore). Depending on the amount, you may need to juggle the gasket thickness or cut the piston tops slightly to achieve zero deck or the desired piston-to-head clearance and optimum amount of quench. If your block had to be cut .010 inch to bring the decks in line with the crank centerline, you would have 0.008-inch positive deck and only 0.031-inch clearance with a 0.039-inch gasket. There is some wiggle room here, but it requires careful attention. If you are ordering custom pistons, you can specify a custom pin height to accommodate any dimensional discrepancies in your block. The piston manufacturer will tell you if your proposed combination will interfere with the ring package and, in the case of minor interference, they can supply a support ring for the bottom ring to ensure stability.

In summary, the stroke length divided by 2 plus the rod length and pin height adds up to the total assembled height within the bore. Their total is compared to the fixed block height to determine final deck height and piston-to-head clearance with the appropriate head gasket and deck milling if required. This formula can help you brainstorm rod and stroke combinations to see how much rod length you can run based on pin height dimensions that safely accommodate your desired piston and ring package.

The assembled deck height (ADH) incorporates half the stroke length, plus the rod length and the pin height. The total equals, or nearly equals, the block height from the centerline of the crankshaft to the deck surface of the block; the difference being the deck height at TDC, either positive or negative depending on the tolerance stack.

Failure to record proper measurements and perform the required math often results in inappropriate mechanical relationships, such as an exhaust valve and piston top occupying the same space at the same time.

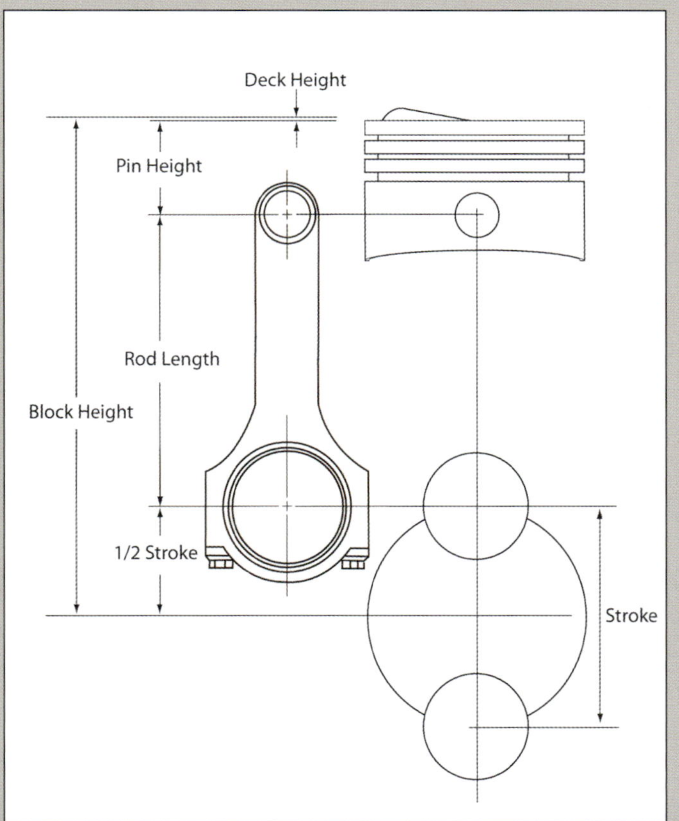

well-known 301-ci hot rod engine of the early 1960s. The factory later duplicated this engine with the 302 small-block for the first-generation Z28 Camaro.

Sonic Checking

Sonic checking is an ultrasonic procedure that provides accurate cylinder-wall-thickness measurements to support bore and stroke calculations. While most factory and, aftermarket race blocks now come with a factory sonic-check sheet, many builders prefer to verify the sheet and in the case of previously bored blocks, it is wise to determine the thickness of the cylinder walls. Most race blocks now provide cylinders with at least 0.250- to 0.300-inch wall thickness and it is important to maintain as much of that as possible. Sonic checking is not a lengthy process and most shops that perform it regularly have their own sheets for recording the numbers. The sonic checker device comes with standards that are used to calibrate the system prior to use. They have a known thickness and are made in a curved shape to simulate the cylinder bores. Some builders break up old blocks and keep curved sections of broken cylinder walls around to use as real-world calibration samples that can be easily measured for comparison.

Once a unit is calibrated, gel is applied to the sensor and the sensor is held firmly against the cylinder wall at specified locations depending on the type of block. Most builders prefer to check the cylinders at four equally spaced locations starting with the primary thrust surface and working their way around the bore 90 degrees at a time about 1½ to 2 inches down from the deck surface. Once these numbers are recorded, they repeat the same process roughly halfway down the bore. Some even record numbers at the bottom of the bore. When the process is completed the builder and/or machinist has an accurate roadmap of the block's cylinder wall thicknesses.

The primary, or major, thrust side is located opposite to the rotation of the engine. For clockwise rotation, stand in front of the engine and face toward it. The major thrust surface is the left side of each bank of cylinders; it's toward the passenger side of the block for every cylinder. That's where you want the thickest readings—typically 0.300 inch or better, but no less than 0.250 inch for racing.

The minor thrust side is the opposite wall, or the right side of all the cylinders as you face the front of the block. If you have counter-clockwise rotation, or, as in some cases, you are building a reverse rotation engine, the major thrust surfaces all shift to the opposite side. Non-thrust surfaces opposite the wrist pin axis in each bore (front and back side of each bore) are typically the thinnest and some builders actually accept walls as thin as 0.100 inch in this area.

Practice Calculations

Grab your calculator and fill in the blanks for the following engines. Calculate the displacement for the first three, the stroke for the second three, and the bore for the last three. Compare your answers to those shown below.

Vehicle	*Bore x Stroke*
1. 1962 Chevy	4.00 x 3.25 = __________ ci
2. 1968 Dodge	4.25 x 3.75 = __________ ci
3. 1957 Ford	3.80 x 3.44 = __________ ci
4. 1970 Plymouth	4.04 x __________ = 340 ci
5. 1966 Pontiac	4.09 x __________ = 421 ci
6. 1951 Ford	3.19 x __________ = 239 ci
7. 1968 Chevy	__________ = x 3.48 = 350 ci
8. 1964 Ford	__________ = x 3.78 = 427 ci
9. 1961 Dodge	__________ = x 3.75 = 413 ci

Answers

1. 327	4. 3.31	7. 4.00
2. 426	5. 4.00	8. 4.23
3. 312	6. 3.75	9. 4.19

Displacement Formulas at a Glance

Pi symbol = π

Pi = 3.1415927

Pi ÷ 4 = 0.7853982 or 0.7854

Cylinder Volume = bore^2 x stroke x 0.7854

Displacement = bore^2 x stroke x 0.7854 x number of cylinders

Stroke = displacement ÷ (bore^2 x 0.7854 x number of cylinders)

Bore = $\sqrt{[\text{displacement} \div (0.7854 \text{ x stroke x number of cylinders})]}$

Bore/Stroke Ratio = bore ÷ stroke

Rod/Stroke Ratio = rod (center to center length) ÷ stroke

Equivalent Displacement = SV x 3

SV = swept volume

Conversions

Conversion	Multiply by
inches to mm	25.4
mm to inches	0.0393701
cc to liters	0.001
liters to cc	1,000
liters to ci	61.023744
ci to liters	0.0163871, or divide by 61.024

COMPRESSION RATIO

Piston top configuration is one of many factors affecting an engine's compression ratio and fuel octane tolerance.

An engine's compression ratio is a big deal. You never see a low-compression racing engine unless it is arbitrarily limited by some class restriction. Higher compression ratios yield more power in racing engines and street engines. Everyone remembers the anemic low-compression 1970s and nobody wants to repeat them. Once the OEMs gained greater control over fuel and spark with EFI and electronic engine management, compression ratios shot up again because automakers know it makes more power and yields higher fuel efficiency. Higher compression ratios are the main reason diesel engines consistently produce better fuel economy than gasoline engines.

Performance applications have to consider compression ratios carefully regardless of whether they are normally aspirated or highly boosted via supercharging. We want all the power and efficiency we can get, but a poor combination of parts can unduly influence an engine's fuel octane tolerance with potentially disastrous results.

It is very important to know or predict compression ratio with a high degree of certainty, so appropriate fuel choices can be made. Now that we have low- and medium-octane gasoline, and high-octane E85 ethanol plus racing fuels, it is more important than ever to match compression ratio to the intended application and the fuel that will be burned. In the case of new engine builds, a suitable mix of components can be tailored to meet a target compression ratio that is either octane friendly or in some cases, sanctioning body mandated.

Octane-limited engines are always at risk for fatal engine damage. That's why engines in the 1980s began sporting knock sensors that would signal the onboard computer to retard the spark advance when the onset of detonation was detected. Today we have the luxury of engine management controls that allow us to run higher compression ratios, but we still have to calculate them according to specific requirements.

Compression ratio is an effective means of limiting power in some racing series. It is also used to curb the cost of many racing venues. It typically influences piston and cylinder head selection where a particular cylinder head may also be specified by a sanctioning body. When cylinder head and chamber size are dictated, piston configuration, deck height, and gasket thickness must be juggled to chase the compression ratio requirement. Short tracks frequently enforce a 9:1 rule while NASCAR engines are limited to 12:1. Unlimited drag racing and Bonneville engines often exceed 14:1, while stock class drag racers are limited to the original factory compression ratio of their particular vehicle.

Compression-ratio limits can be useful to a degree since they generally dictate flat-top pistons, which encourage efficient combustion while maintaining desirable quench to promote charge turbulence and maintain mixture quality. Hypereutectic pistons are often specified, although forgings are permitted in some series. There is certainly less bang for the buck without higher compression ratios but, given specific parameters, experienced engine builders adjust contributing components to best suit any fixed compression ratio, particularly with an eye toward increasing the effective compression ratio via appropriate camshaft timing and effective inlet tuning.

Factors Affecting Compression Ratio

Quickly name ten or more things that affect or are affected by compression ratio. If you can't, consider the following:

- Fuel octane
- Fuel mixture quality (droplet size)
- Cylinder volume
- Combustion chamber volume
- Deck height
- Compressed gasket thickness
- Gasket shape
- Piston to head clearance
- Quench area
- Dome or dome volume
- Dish volume
- Ignition timing
- Valve relief volume
- Crevice volume
- Bore chamfer

The formula for calculating compression ratio is pretty simple. We'll work with some examples in a moment, but first let's examine the influence of the items on our list, particularly those within our control during an engine assembly process. Of course fuel octane tolerance is a primary concern so we need to know what fuel we will be using. The mixture quality of that fuel is largely determined by air temperature, fuel blend, and the induction components that meter fuel to the engine. These would include the carburetor or the fuel injectors, the intake manifold, cylinder heads, and valves. Even the camshaft timing can have an effect on dynamic compression—or cylinder pressure. Those are all things we can control and so are the items on our list, all of which are right there inside the cylinder exerting their influence on compression ratio. Consider the basic formula.

$$\text{Compression Ratio (CR)} = (V_1 + V_2) \div V_2$$

Where:
V_1 = cylinder volume
V_2 = combustion chamber volume

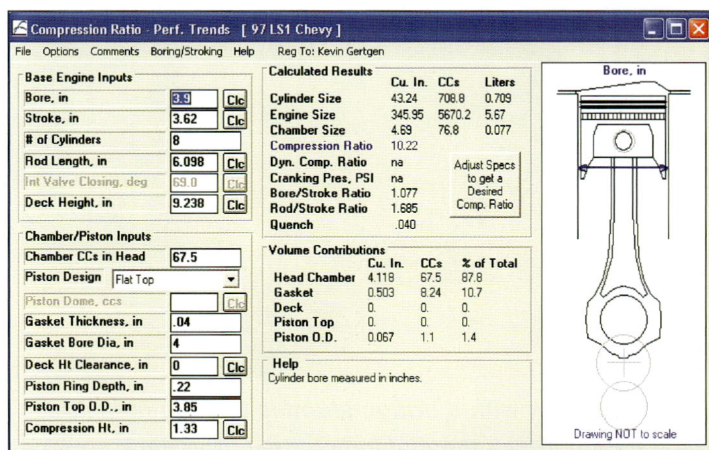

Performance Trends' Compression Ratio Calculator is a robust tool that incorporates all measured and calculated components of the compression ratio formula to provide precise compression ratio calculations.

In practice, V_2 is actually called the clearance volume or the compression volume because it incorporates all of the items on our list and actually represents the total combustion space above the piston. This is the space the cylinder volume gets squeezed into during compression. I will call it the compression volume for our discussion. So the formula actually establishes the ratio between the total cylinder volume with the piston at the bottom of its stroke to the volume of the cylinder with the piston at the top of its stroke. Each item on our list changes the value of V_2 to some degree or another and that has a profound influence on the actual working compression ratio.

Most head gaskets have a multilayered construction and all the best ones provide a published compressed thickness and volume. If your gasket volume isn't known, you can still measure it as detailed in the accompanying text.

Deck Height

There are two types of deck height: positive and negative. On most engines the piston stops slightly below the block deck surface when it is at TDC, sometimes 0.020 inch or more. This is called positive deck height because the block deck is still above the piston top. No matter how small it might be, this distance contributes additional volume to the combustion space V_2 above the piston. This volume must be calculated and added to V_2. In some cases the piston protrudes slightly out of the bore. This is called negative deck height and its volume must be subtracted from V_2 because it subtracts volume from the combustion space.

Compressed Gasket Thickness

Head gasket volume also adds to the compression volume. It is dictated by the compressed gasket thickness, the gasket bore diameter, and the gasket shape. Many head gaskets are slightly larger than the cylinder bore diameter and they often have irregular shapes. Deck height and gasket thickness also influence piston-to-head clearance which must be considered, especially in high-RPM applications.

Steel connecting rods don't really stretch, so you can get that piston right up close to the cylinder head (without consequence to improve quench). Quench is where the flat top portion of the piston rises very close to the head, which

A dial indicator with a bridge stand is used to measure deck height. Place the dial indictor over the deck surface and zero the dial. Then rotate the piston to TDC and measure the difference to the piston top. Take your measurement along the piston pin axis to get the mean deck height.

Domed pistons raise compression by displacing volume in the combustion space above the piston deck, but shallow combustion chambers are the current trend for raising compression ratio. By eliminating or reducing the dome, combustion efficiency is improved because the dome is not there to block the flame kernel that initiates at the spark plug.

Flat tops are the most common piston configuration. To a degree they simplify compression ratio calculations, but you still have to deal with the valve reliefs. They promote superior combustion with good quench and turbulence properties.

tends to force or squirt the charge toward the spark plug with high chamber turbulence to improve the burn.

Aluminum connecting rods have some degree of elasticity, so they require increased piston-to-head clearance to avoid physical contact and the ensuing damage at high engine speeds.

These requirements can influence your choice of gasket thickness and thus compression ratio. Often you have to juggle the combination to get what you want. Calculating it ahead of time helps you make those choices.

Dome Volume and Dish Volume

If the piston has a raised dome to increase compression, the dome's volume must be considered in the compression ratio calculation. Dome volume must be subtracted from V_2 since it reduces compression volume. Dish volume is added to V_2 since it adds volume. And while you're figuring dome and dish volumes, you also have to account for the volume of any valve reliefs in the piston top.

And if you really want to pick nits, you can include the crevice volume above the top piston ring and the chamfer volume at the top of the cylinder bore. While

Dished pistons are designed to reduce compression ratio by increasing the compression volume above the piston. Many of them don't have valve reliefs because the dish is already deep enough. You can use the published dish volume for your compression ratio calculations or cc a piston top to verify it.

infinitely small, they still contribute to the total volume of V_2 in the equation. Crevice volume is the tiny space between the piston and the cylinder wall above the top ring. Typically only a few thousandths of an inch, it is still multiplied by the bore circumference and has a volumetric value. And if the cylinder bore also has a large

This comparison between a domed piston and a dished piston illustrates how the dome projects into the combustion chamber to raise compression by reducing chamber volume while the dished piston increases combustion space volume to reduce compression ratio.

chamfer to aid piston installation, it also contributes volume to the combustion space. Crazy huh?

Some of these volumes are inconsequential in most cases, but you should know about them so you can decide whether to incorporate them in your calculations. If you are building a high-performance engine, you'll be repeatedly measuring and altering many of these volumes during pre-assembly mockups. Proper clearance between rapidly moving parts is essential and unforgiving so you have to establish them first. An awareness of their effect on compression ratio helps you consider your alterations and parts selection accordingly.

Finding V_2

Compression ratio is no simple thing, especially when you break it down into all its contributing factors. Still, it's manageable and there are many ways to look at it. While this is primarily an engine math book, it is still important to comprehend all the factors and the manner in which they affect engine performance. Compression ratio is simply a measure of how tightly the incoming charge is squeezed before the spark plug ignites it. It is created by the combined volume of the cylinder and the compression volume when the piston reaches TDC. In reality it is controlled by the swept volume of the cylinder and any combination of the various combustion space volumes that make up the compression volume V_2. Since that is where all the variables are, that is where you have to concentrate your efforts to achieve the compression ratio you want.

To see how much influence these factors have, let's compare the basic formula to the same formula with all factors considered. As previously discussed, the various contributing factors are either additive or subtractive from the total compression volume. The combustion chamber is the primary value. All other volumes are either added to it or subtracted from it prior to working the basic equation.

$$CR = V_1 + V_2 \div V_2$$

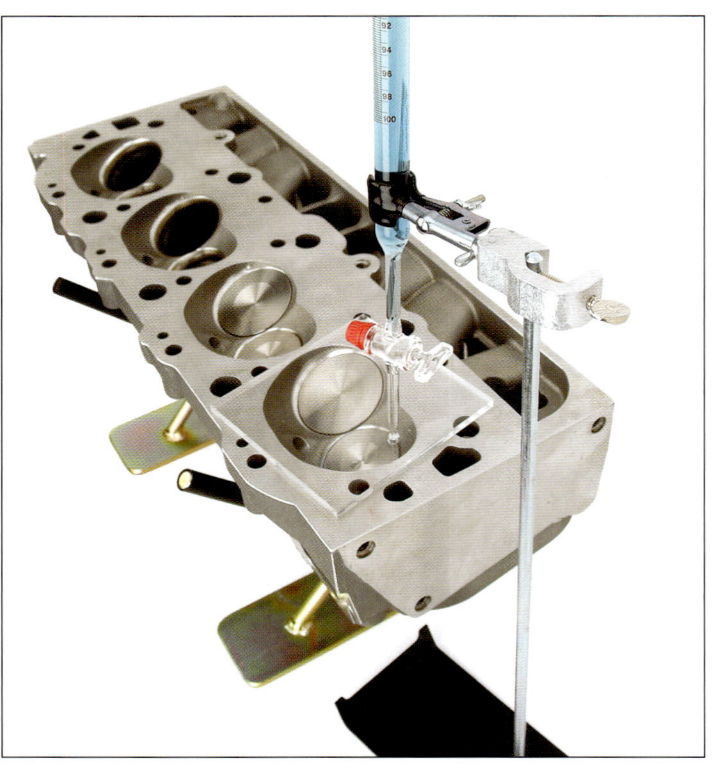

Determine combustion chamber volume by filling the chamber with water or alcohol from a graduated burette calibrated in cubic centimeters (cc). Tighten a spark plug in the chamber with both valves installed. Then use a light grease to seal the deck surface. Place the plastic cc plate over the chamber and position the head so the fill hole is at the highest point. Fill the chamber and read off the burette. Divide by 16.4 to convert to cubic inches.

Note that V_1 is constant, but V_2 can vary by a considerable degree when you start adding and subtracting the various values that contribute to it. In the simple formula, V_2 is called chamber volume, but we know it is really compression volume because it incorporates other factors. If you add all the other factors, it makes a very long equation. You can break it down by calculating absolute V_2 before plugging it into the equation. This requires precise measurements although, in practice, published values for gasket volume, dome and dish volume and valve relief volumes are often substituted. Crevice volume and chamfer volume are typically ignored because they are so small. The following list is called the V_2 stack.

To find absolute V_2, begin with the measured chamber volume with cubic centimeters converted to cubic inches, then:

add deck volume (or subtract if deck is negative)
add compressed gasket volume
add dish volume (or subtract if dome)
subtract dome volume (or add if dish)
add valve relief volume
add crevice volume (if desired)
add chamfer volume (if desired)

This is simple, but somewhat tedious to measure and calculate so many engine builders choose to measure it all at once by cc'ing a cylinder with a piston in it. I will explain how to do that in a moment, but first let's discuss how to determine all the individual volumes that make up V_2.

Deck Volume

Calculate deck volume as if it were a very short cylinder. The positive or negative deck measurement represents the height dimension in the formula which uses the displacement constant 0.7854.

Example: for a 0.020-inch-positive deck height on a 4-inch bore

$$4^2 \times 0.020 \times 0.7854 = 0.251328 \text{ ci}$$

This will be added to the V_2 stack because it increases the compression volume. If the deck measurement were negative (piston above deck), the result would be subtracted from the V_2 stack because it reduces compression volume. An interesting fact is that all small-block Chevys are positive-deck engines, but the newer Gen III engines all have negative decks.

Chamber Volume

Combustion chamber volume is measured directly by cc'ing a chamber with a graduated burette. Note that chamber size in cubic centimeters must be converted to cubic inches. Divide by 16.4 to make the conversion. This will be your base volume for calculating compression ratio. All other relevant volumes are either added to or subtracted from the chamber volume to determine the compression volume.

To cc a cylinder, coat the cylinder wall with light grease or oil to help seal the right gap. Rotate the engine until the piston top is far enough down the bore to clear the dome. Measure the depth with a dial indictor and compute the empty volume using the cylinder volume formula. Then cc the cylinder to find out how much volume the dome displaces. Subtract that value from your compression volume.

Gasket Volume

In most cases gasket volume is published by the gasket manufacturer and it is safe to add (+) to the V_2 stack. When no published number is available, builders often fudge by calculating the volume based on a perfect circle (just like deck height volume). The problem is that the gasket bore diameter is often larger than the cylinder bore diameter and frequently irregular in shape. If it is perfectly round, you can calculate using the cylinder volume formula with the appropriate diameter and compressed thickness.

If the shape is irregular, you can fudge it or use the string-and-tape method to find the true circumference of the gasket bore and then treat it as a perfect circle for the calculation. Tape the gasket to a flat surface and use small bits of tape to secure a thin string around the perimeter of the gasket bore opening. Once you reach the starting point, carefully cut the string and measure its length.

Using the formula for the circumference of a circle, you can find the appropriate diameter to use in the gasket volume calculation. Suppose you have a 4-inch cylinder bore, and your gasket bore is noticeably larger with an irregular D-shape around the valves (which is typical of many head gaskets). You carefully string the perimeter and get a length of 13¹⁄₁₆ inches. Convert to decimals and you have 13.0625 inches. Now plug this dimension into the formula.

$$\text{Circumference} = 2\,\pi\,r, \text{ or } C = \pi\,d$$

Where:
r = radius
d = diameter

$$d = C \div \pi$$

$$13.0625 \div 3.14 = 4.16 \text{ inches}$$

That's your true gasket bore diameter and it can now be plugged into the gasket volume formula:

$$\text{True Gasket Volume} =$$
$$4.16^2 \times \text{gasket thickness} \times 0.7854$$

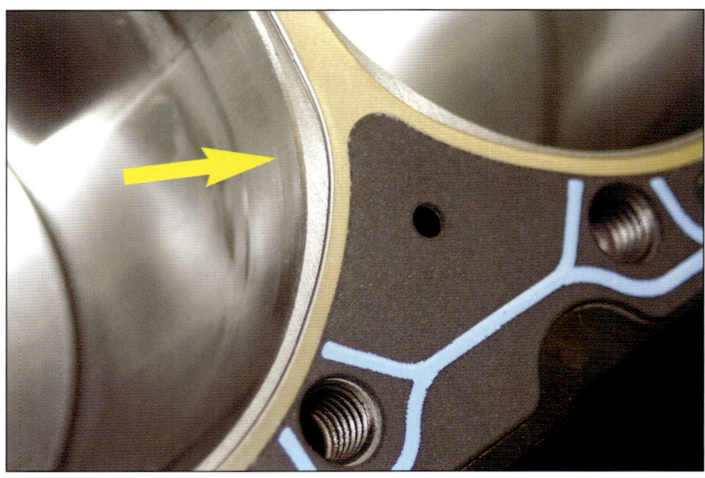

This is an example of a head gasket with an irregular shape and a diameter that is larger than the bore. You'll usually find an eyebrow like this adjacent to both valves. This must be included in your compression ratio calculation. You can string the perimeter of an irregular gasket and use the string length to compute the gasket volume based on the measured thickness (see text).

Dish Volume

Dish volumes are typically published so you can usually plug them right into your V_2 stack. But let's say your block has already been decked a couple of times and it's a little shorter than normal so the piston has a negative deck by some amount that is more than what you are comfortable with for piston-to-head clearance.

Most pistons tolerate some degree of piston deck shaving (up to 0.100 inch or even more in many cases) so you decide to cut them down to achieve a zero deck (piston flush with block deck surface). This is easy to do with flat-top and dished pistons; it's slightly more complicated with dome pistons (rarely done).

If your piston is dished and you cut it down by some amount, you can cc the dish and add the new volume to your V_2 stack. Or you can use the cylinder volume formula to calculate the difference if you have an accurate measurement of depth and diameter. In practice this is never easy because the dish is not always perfectly round and is often D-shaped and curved at the bottom.

Dome Volume

Dome volumes are also published by piston manufacturers. They are pretty accurate so you can safely subtract that volume from your V_2 stack as long as you haven't modified the dome by fitting it to the chamber shape, cutting deeper valve reliefs or cutting a fire slot for the spark plug. Sometimes during mockup assembly you identify a small spot where the piston dome contacts the chamber roof during rotation. These spots are typically cut down to achieve the minimum clearance and that alters the dome volume, which then requires you to measure it. Moroso sells a simple tool for cc'ing dome volumes and it comes in handy in this situation. Remember, dome volume gets subtracted from you final V_2 stack.

Valve Reliefs

Valve reliefs are easy enough to cc on a flat-top piston and most manufacturers already publish the volumes for all their pistons. Here again, you only need to measure if you have significantly plunge cut the reliefs to gain adequate piston-to-valve clearance. Whatever the volume, it is an additive value to your V_2 stack.

Crevice Volume

Crevice volumes are minimal and are not frequently incorporated in compression ratio calculations, but some builders find reason to do so. Some are simply detail freaks. Crevice volumes have long been known to affect emissions because they provide a hiding place for small amounts of fuel mixture that didn't quite participate in the combustion process. This is mostly important to chemists and combustion engineers but if you want to include it, here's how.

$$CV = (d_1 - d_2) \times c \times r$$

Where:
d_1 = bore diameter
d_2 = piston diameter at top ring land
c = bore circumference
r = top ring depth from piston deck

So, with a 4.00-inch bore, 0.010-inch piston-to-wall clearance above the top ring and the ring 0.125-inch down the bore we calculate:

$$CV = (4.00 - 3.990) \times 12.56 \times 0.125 = 0.0157 \text{ ci}$$

12.56 is the circumference of the bore and is found by multiplying bore diameter times pi. If you want to be precise, add the result of your final calculation to your V_2 stack.

Chamfer Volume

Most machinists put a chamfer at the top of the bore to help guide the rings into the bore during assembly. Sometimes it is quite substantial so you may want to include it in your calculations. Chamfers are generally 40

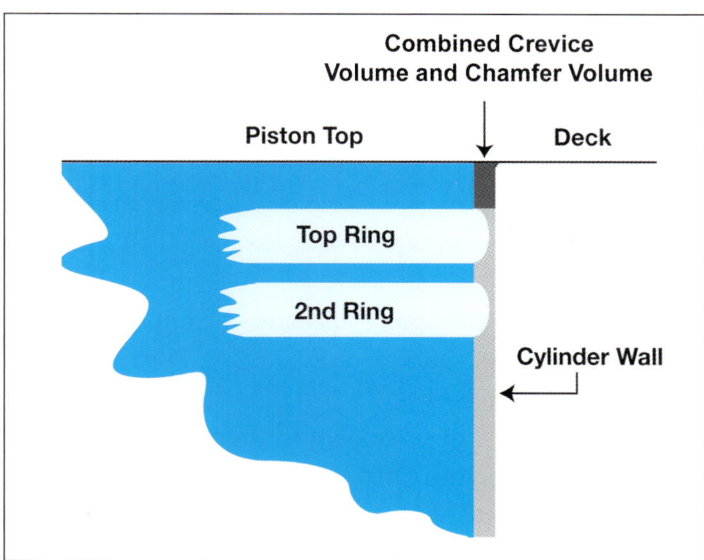

The combined crevice and chamfer volume is the space between the cylinder wall and the piston above the top piston ring. It is shown here by the dark shaded area above the ring.

A large chamfer at the top of the bore also contributes to compression volume in some small amount, but not enough to concern most builders. If compression volume is determined by cc'ing a cylinder, both the crevice volume and the chamfer volume are included in the measurement.

to 60 degrees and even at these small dimensions, you can pretty much treat them as squares or rectangles when viewed end on. Use the same formula as for crevice volume, but start with the larger outside dimension where the chamfer starts (see fig. 1, page 35)

If it is about 0.060 larger than the cylinder bore:

$$CV = [(4.060 - 4.000) \times 12.748 \times 0.060] \div 2 = 0.022 \text{ ci}$$

Note that the "c" dimension has changed because we now have a 4.06-inch outer diameter (4.06 x 3.14 = 12.748). The depth is only 0.060 inch and we have to divide the result by 2 to complete the formula for the area of a triangle and thus the volume when the length is added.

The result is more than the crevice volume, but still nothing substantial, which is why most engine builders exclude crevice volume and chamfer volume from their calculations. If you use them, remember that they are additive and thus are added to your V_2 stack. Crevice volume and chamber volume partially occupy the same space, but it is more convenient to calculate them separately.

Now let's review our V_2 stack with calculated values based on the following dimensions:

V_1
Bore/Stroke, 4.00 x 3.00 inches ....37.699 ci

V_2
Chamber volume, 64 cc..............................3.902 ci
Deck height, 0.020 positive ....0.251 ci V_2 +
Gasket thickness, 0.015 (published) ....0.194 ci V_2 +
Flat top (or dish/dome) ....0.000 (flat) ±
Valve relief, 4 cc (published) ....0.243 ci V_2 +
Crevice volume, calculated ....0.015 ci V_2 +
Chamfer volume, calculated ....0.022 ci V_2 +
Total 4.627 ci = V_2

$$V_1 + V_2 \div V_2 = CR$$

$$(37.699 + 4.627) \div 4.627 = 9.14 \text{ CR}$$

That's adequate, but perhaps a little low for street performance. If you zero deck the block and eliminate the deck height dimension from V_2, you can raise the

compression ratio to 9.61:1, just about right for a street engine. That small change illustrates just how much influence all the small volumes that make up V_2 have on the final compression ratio.

Displacement Ratio

The concept of displacement ratio is not frequently used, but it should be understood because it can sometimes help us estimate the amount of combustion chamber milling that will achieve a desired compression ratio. As we have seen, compression ratio is the combined volume of the swept cylinder volume and the compression volume divided by the compression volume (see sidebar, page 37). Displacement ratio is simply the swept cylinder volume divided by the compression volume:

$$\text{Compression ratio} = V_1 + V_2 \div V_2$$

$$\text{Displacement ratio} = V_1 \div V_2$$

Note that the compression ratio is always 1 greater than the displacement ratio. By rearranging the displacement ratio formula we can calculate a new compression volume V_2 that will yield the desired compression ratio.

$$\text{New } V_2 = V_1 \div \text{displacement ratio}$$

Now we can derive a formula for cylinder head milling:

$$\text{Mill Cut} =$$
$$[(\text{new displacement ratio} - \text{old displacement ratio}) \div$$
$$(\text{new displacement ratio} \times \text{old displacement ratio})] \times \text{stroke}$$

Recall that we previously calculated a compression ratio of 9.14:1 for a 4.00-inch bore and a 3-inch stroke. Since displacement ratio is always 1 less than compression ratio, we use 8.14 for the displacement ratio in our formula. We already saw that eliminating 0.020 inch of deck height raised compression to 9.61:1. Now let's see what reducing the combustion volume does. Since we want to raise the compression to 9.61:1, our displacement ratio is 8.61.

$$\text{Mill Cut} =$$
$$[(8.61 - 8.14) \div (8.61 \times 8.14)] \times 3 = 0.0201 \text{ inch}$$

Compression and Displacement Ratios

The illustration shows the relationships of the various volumes that make up the compression ratio and the displacement ratio. To simplify the illustration, we're assuming a swept volume of 100 ci in the cylinder (V_1) and 10 ci of compression volume in the combustion space above the piston at TDC (V_2). Study the illustration to see the difference in the two formulas.

$$\text{Compression Ratio} = (V_1 + V_2) \div V_2$$

$$\text{Compression Ratio} = (90 + 10 \div 10) = 10{:}1 \text{ CR}$$

or

$$\text{Displacement Ratio} = V_1 \div V_2$$

$$\text{Displacement Ratio} = 90 \div 10 = 9{:}1 \text{ DR}$$

Swept volume is dictated by the bore diameter and stroke length. The compression volume is usually called the chamber volume, but it actually incorporates all the smaller volumes that contribute to it. In addition to the combustion chamber it includes the deck volume, gasket volume, dome or dish volumes, valve reliefs, and even the crevice volume and the chamfer volume.

The swept volume of a cylinder is calculated with the formula for cylinder volume.

$$\text{Swept Volume, or } V_1 = \text{bore}^2 \text{ x stroke x } 0.7854$$

Compression volume is determined by building the V_2 stack, a combination of all the smaller volumes that make up the combustion space above the piston at TDC. This incorporates both measured and calculated values. To compile a V_2 stack you need all or most of the following:

- Chamber volume converted to cubic inches (divide by 16.4)
- Compressed gasket volume
- Deck height volume
- Dome or dish volume (if present)
- Valve relief volume

All of these volumes are added together to determine the true compression volume (V_2). The one exception is dome volume which reduces total volume so it is always subtracted. Two other values, crevice volume and chamfer volume, may also be considered but their values are typically inconsequential so most people leave them out.

To understand the compression ratio imagine the lightly shaded area in the cylinder on the left being compressed into the darker shaded area at the top of the cylinder on the right.

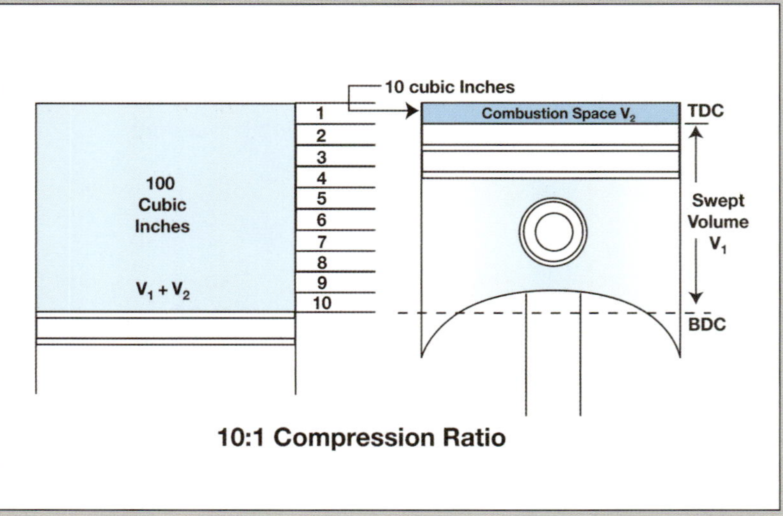

10:1 Compression Ratio

How To CC a Cylinder

Cc'ing a cylinder allows you to account for all the compression ratio variables in one easy operation. Cylinder cc'ing provides you with the exact displacement volume for a domed piston or the volume increase if using a dished piston. It requires the same equipment used to cc a combustion chamber, but in this case you are actually determining the volume of a fixed cylinder dimension to the volume of that same cylinder with the actual piston top configuration that contributes to the final compression ratio.

Install the piston and rod assembly in the block with the piston positioned at the bottom of the bore. If the piston has a tall dome, determine the depth required to keep the dome below the deck surface during the measurement. Coat the cylinder wall with light grease at the approximate depth that you plan to raise the piston.

With the crank installed in the block, mock up a piston and rod assembly with the proper bearings and a dummy top-piston ring. If you work carefully you can use the actual ring you plan to install. The ring is necessary to help seal the cylinder during the cc'ing procedure.

Slowly rotate the crank and bring the piston to your exact predetermined depth. This is the depth from the deck surface of the block to the flat portion of the piston deck as measured with a dial indicator or a depth micrometer.

Lightly coat the deck surface with grease and press your Plexiglas or glass cc'ing plate down over the bore with the filling hole located toward the side of the cylinder closest to the lifter gallery. Rotate the block on the engine stand so that the filling hole is positioned at the highest point.

With your burette filled to the zero mark, position it over the filling hole, open the petcock, and begin filling the cylinder with checking fluid. Note that in this case you will likely have to stop and refill the burette one or more times because you are dealing with a considerably larger volume than the combustion chamber. When refilling the burette take care to fill it exactly to the zero mark each time.

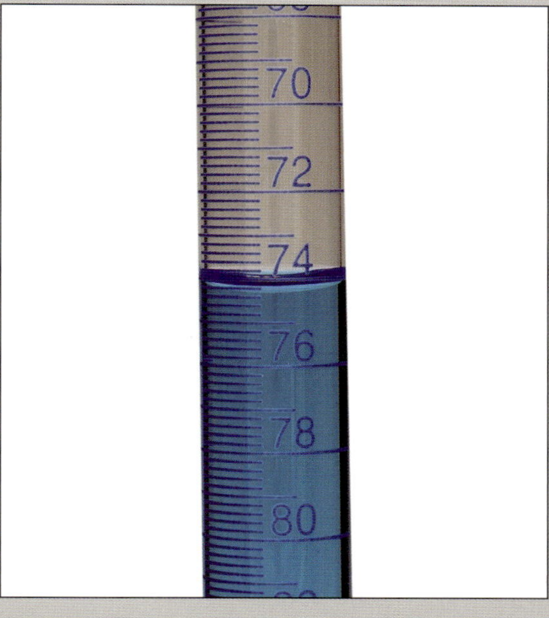

When the fluid reaches the filling hole, close the petcock and read the burette. Most burettes have a 100-cc capacity. If, for example, you have refilled the burette twice and the burette reads 74 cc on the third filling sequence, you would add 100 cc plus 100 cc plus 74 cc to get the total. Once the measurement is completed you add or subtract it from your V_2 value to determine the exact compression ratio with all factors incorporated.

Critical Bore and Stroke Relationships

The Illustration shows the piston at top dead center (TDC) in its travel within the cylinder bore. All of the various contributors to compression volume are located here. The deck height notes how far the piston top stops below the block deck at the top of its travel. The compressed gasket thickness shows the location of the gasket between the block and the cylinder head. Note how the deck height and gasket thickness contribute additional volume to the combustion space formed by the combustion chamber. Also note that the gasket diameter is larger than the cylinder bore and that its shape is irregular because it is farther away from the bore on one side than the other. This is not always the case, but it must always be accounted for, either way.

You cannot see a piston dish in this side view, but the dotted line indicates a possible dome that would require measurement if present. Also note the slight chamfer at the top of the cylinder bore and the crevice volume, which is shaded in light blue on either side of the piston above the top ring. This view also shows an exaggerated view of the quench area where the flat top of the piston almost touches the cylinder head. In most cases this quench area is little more than the combined dimension of the deck height and the gasket thickness.

Everything here contributes to the combustion volume and, thus, the V_2 stack in the equation. As the piston travels down the bore the length of its stroke, the volume it displaces is the cylinder volume, or V_1 in the compression ratio formula.

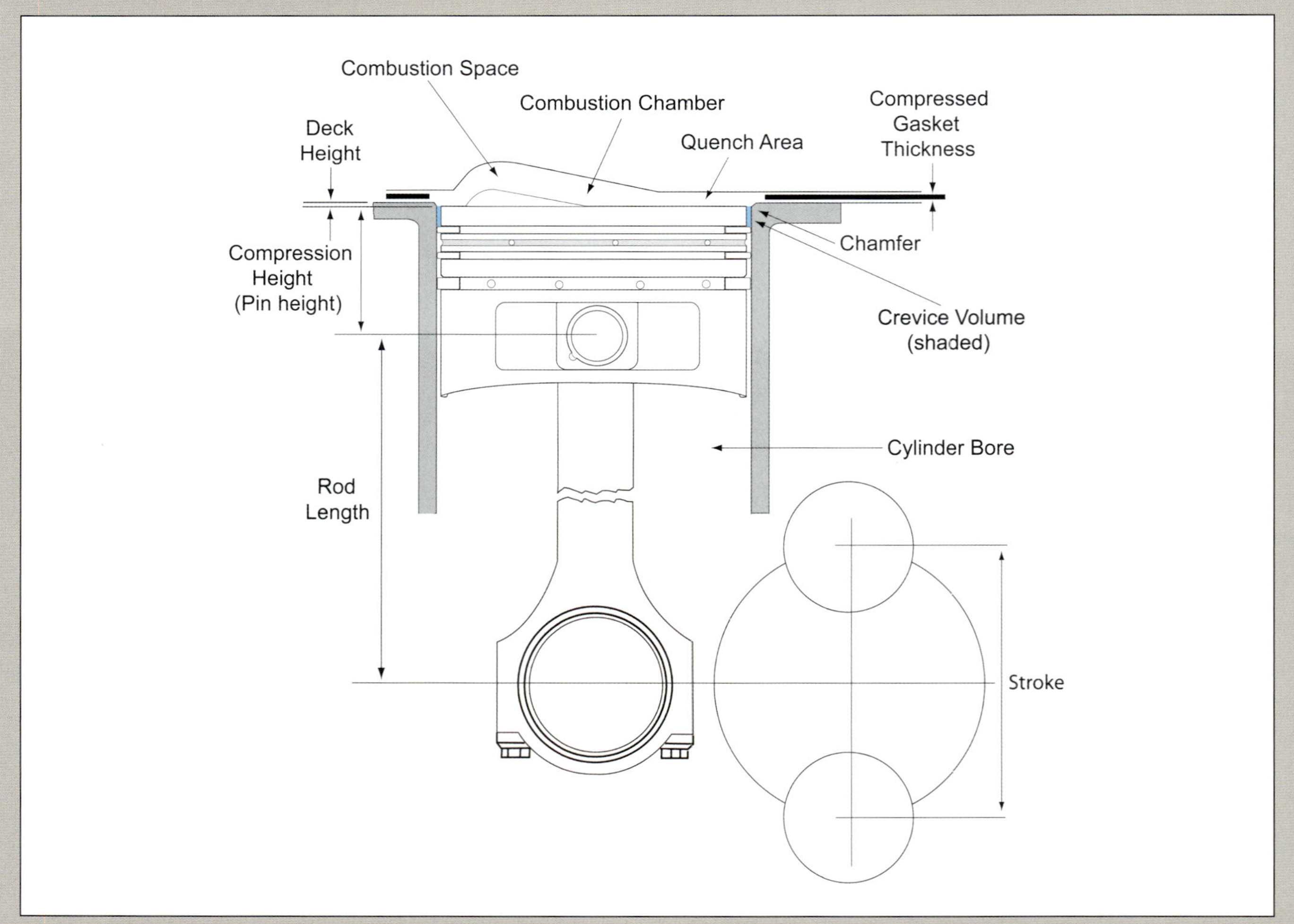

That is almost exactly the same amount as the deck height we eliminated in our previous calculations, but is it correct? Not exactly. In removing the deck height dimension, we accounted for the entire bore diameter of the cylinder. But the D-shaped combustion chamber on our small-block Chevy is only about half the size of the bore. We have to make a deeper cut to get the same result. In this case, about 0.040 inch gives us the desired result. We have to double the cut because we're only dealing with half the area. These are relatively straightforward procedures, but you have to think carefully about them to avoid costly mistakes.

Cranking Compression

Cranking compression is often confused with compression ratio. While compression ratio is a relationship of volumes within a cylinder, cranking compression is actually a measured cylinder pressure taken at the spark plug hole while the engine is bring cranked with the throttle plates held open. The coil wire is removed during this operation to prevent the other cylinders from firing. Cranking compression is the peak pressure achieved within the cylinder during cranking. Higher compression ratios can affect cranking compression, but the two are not related.

Cranking compression is used as an indicator of engine condition and also of the relationship of intake and exhaust valve opening and closing points. Depending upon the condition of the piston rings and valves, a healthy engine typically has a cranking compression between 150 and 180 psi. A good performance engine can easily have a cranking compression of more than 200 psi. Some are a little higher and a few are much lower. The important thing is that all cylinders should read the same during a compression test. A low reading on any cylinder typically indicates leaky valves or piston rings. Big camshafts with a lot of valve overlap can also affect cranking compression but not by any great amount. As long as all the cylinders match within 5 or 10 psi, you probably have a healthy engine. Inexpensive compression gauges are available at any auto parts store.

Compression Ratio Formulas at a Glance

Compression Ratio (CR) = $(V_1 + V_2) \div V_2$

Circumference = $2\pi r$, or $C = \pi d$

Chamber Volume = Measured cc $\div$ 16.4

Crevice Volume = $(d_1 - d_2)$ x c x r

Chamfer Volume = $[(d_1 - d_2)$ x c x r$] \div 2$

Displacement ratio = $V_1 \div V_2$

Mill Cut =

[(new displacement ratio − old displacement ratio) $\div$

(new displacement ratio x old displacement ratio)] x stroke

Swept Volume, or V_1 = $bore^2$ x stroke x 0.7854

PISTON SPEED

Pistons come in all shapes, sizes and alloys. Calculating the piston speed for any given application is essential to safe engine operation.

Piston speed generally refers to the average or mean speed of the piston as it moves up and down in the cylinder bore during each crankshaft revolution. Since the piston actually comes to a complete stop at the top of the stroke (TDC) and at the bottom of the stroke (BDC), its speed and acceleration at any given point is always changing. The piston is always accelerating from or decelerating to zero speed. The formula for mean piston speed yields an average speed based on two times the stroke (up and down for one revolution), times the engine speed (RPM) divided by 12 to convert to feet per minute (fpm). To simplify the formula, divide the numerator and the denominator by 2.

$$\text{Piston Speed}_{fpm} = \text{stroke} \times \text{RPM} \div 6$$

Let's work an example for a 302 Ford that has a stroke of 3 inches and a maximum engine speed of 6,000 rpm.

$$\text{Piston Speed}_{fpm} = 3 \times 6{,}000 \div 6 = 3{,}000 \text{ fpm}$$

Note that if you did not simplify the formula, the answer still comes out the same. It is the distance the piston travels in one minute. To convert the answer to MPH, multiply the answer by 60 to get feet per hour. Then divide by 5,280 to get miles per hour.

$$\text{MPH} = (\text{mean speed} \times 60) \div \text{feet per mile}$$
$$\text{MPH} = (3{,}000 \times 60) \div 5{,}280 = 34.09 \text{ mph}$$

Maximum Piston Speed

You can get a very close approximation of maximum piston speed (ignoring rod center-to-center length and rod angularity) with the following formula. Multiply the stroke by pi and divide by 12 to get feet per revolution. Then multiply by the maximum engine speed to get the maximum feet per minute. This speed occurs about mid stroke, where the connecting rod is ninety degrees to the crankpin. Before that point the piston is

A stock cast piston in a 350 Chevy revs to 6,000 rpm just fine. That's 3,480 fpm so the commonly accepted limit of 3,500 fpm is pretty safe. Start revving those cast pistons to 7,000 rpm and you'll be knocking the skirts off of them in short order.

Forged pistons can take the abuse so you're good up to about 5,500 fpm and more, especially in short bursts such as drag racing.

Hypereutectic pistons have higher silicon content and are approximately one third stronger than regular cast pistons. They are used in many sportsman series, but are not recommended for extended use above 6,000 rpm.

Maximum Piston Speed = MPS

$$MPS_{fpm} = (stroke \times \pi \div 12) \times RPM$$

$$MPS_{fpm} = (3.00 \times 3.14 \div 12) \times 6,000$$

$$MPS_{fpm} = (9.42 \div 12) \times 6,000 =$$
$$.785 \times 6,000 = 4,710 \text{ fpm}$$

or
$$MPS_{fpm} \times 60 \div 5,280 = MPH$$

$$4,710 \times 60 \div 5,280 = 53.52 \text{ mph}$$

Mean piston speed has long been used as an indicator of component durability under severe service. It is a good rule for evaluating engine potential and it is even more instructive if you calculate maximum piston speed since one of the axioms of engine performance dictates that power comes from engine speed. The more power strokes per minute, the more power available to do work.

Consider the 2000 Ferrari Formula 1 engine that has a stroke of 1.6-inch. At 18,000 rpm that's a mean piston speed of 4,800 fpm and a peak speed over 7,536 fpm, or more than 85 mph at mid-stroke. Potential disaster for a big block Chevy, but the Formula 1 engine enjoys the luxury of very exotic materials and very small pistons that have much less mass to accelerate.

accelerating; after it the piston is decelerating. When the piston is exactly at either top or bottom dead center it is stopped and there is no acceleration. Using the formula for mean piston speed we calculated 3,000 fpm or 34.09 mph for our 302 Ford. Now let's find the maximum piston speed at 6,000 rpm.

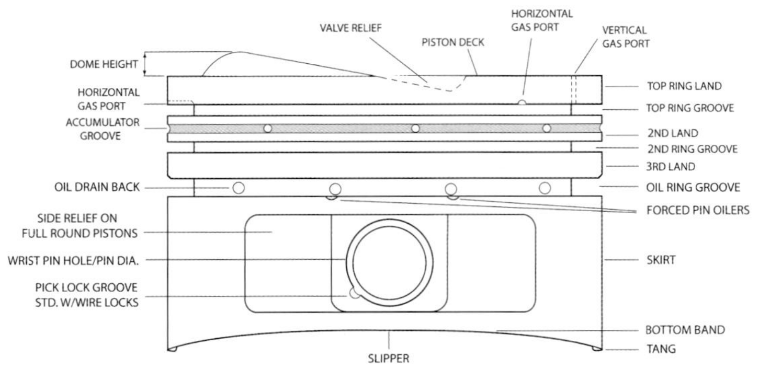

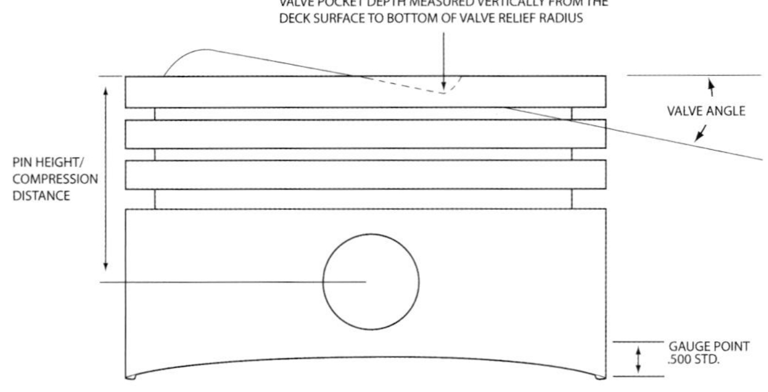

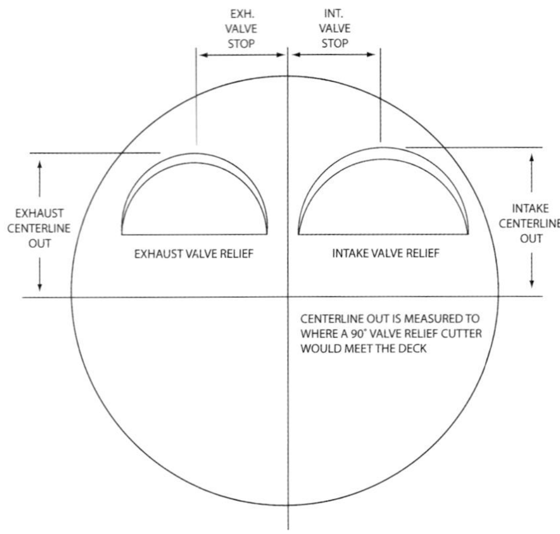

The piston nomenclature diagram illustrates all the important piston features and dimensions. Note in particular the piston pin height, or compression height, which changes according to connecting rod length.

Piston Acceleration

The most important consideration is the instantaneous piston acceleration and the staggering loads placed on the piston, piston pin, and connecting rods and rod bolts. These are the most highly stressed components in the engine. Since an engine's ability to make power is closely tied to the RPM it can turn, every effort is made to lighten valve train components to combat valve float. But the real limit turns out to be piston mass and piston acceleration. A typical 350 Chevy piston weighs 1.3 to 1.6 pounds. Special racing pistons weigh in at less than a pound, but imagine trying to accelerate one to 6,800 fpm (350 Chevy at 7,500 rpm) max piston speed at mid-stroke and then slam it to a dead stop and reverse direction in about 1¾ inches (stroke/2). At TDC the piston is headed for the moon and the rod has to stop it and yank it back the other way. That's enough to pull the piston pin right out of the piston and it does on occasion. It also exerts similar loads on rod bolts and rod caps. Since acceleration (load) is greatest just after TDC, lets calculate the maximum acceleration of a 350 Chevy piston at 7,500 rpm using the stock stroke of 3.48 inches and the stock rod length of 5.7 inches. The formula is:

Max Acceleration =
(rpm^2 x stroke ÷ 2,189) x 1⅓

or

$$MA = (7{,}500^2 \text{ x } 3.48) \div 2{,}189 \text{ x } 1\tfrac{1}{3}$$

$$= 195{,}750{,}000 \div 2{,}189 \text{ x } 1\tfrac{1}{3}$$

$$= 89{,}424.39 \text{ x } 1.3333333$$

$$= 119{,}232.5 \text{ ft/sec}^2$$

That's insane acceleration for a 1.5-pound object that is not a cannon projectile. And because they are captured by the ring grooves, the piston rings are along for the ride; slamming up and down within the ring grooves trying desperately to maintain a seal with the cylinder wall. Is it any wonder that they experience ring flutter at very high engine speeds? Hence the practice of using the tightest ring grooves possible without seizure and the thinnest and lightest rings that have minimal inertia. Try working that formula on the previously mentioned Formula 1 engine and you'll find that the instantaneous acceleration is far beyond the normally accepted limit of 150,000 ft/sec^2. How do they do it? The pistons are changing direction more than 150 times per second. It seems far beyond the physical limitation of the components involved, but then Formula 1 uses some very light and very strong exotic materials.

Fascinating stuff, but somewhat beyond most of our typical applications so we're better off calculating some real-world piston speeds and learning how to rearrange the formula to calculate RPM limits based on an arbitrary piston speed that we don't want to exceed. Consider the following examples when solved for mean piston speed, the calculation you will use most frequently.

302-ci Chevy with 3-inch stroke at 7,000 rpm

Piston Speed = 3 x 7,000 ÷ 6 = 21,000 ÷ 6 = 3,500 fpm

The generally accepted limit for non-race performance applications is about 3,500 to 4,000 fpm, so the 302 is pretty happy at 3,000 fpm, especially since it came equipped with forged pistons.

350-ci Chevy with 3.48-inch stroke at 7,000 rpm
Piston Speed = 3.75 x 7,000 ÷ 6 = 26,250 ÷ 6 = 4,375 fpm

Piston speed increases with RPM, but note that for a fixed RPM, piston speed always increases with stroke. That's one reason why high-RPM engines generally trend toward shorter strokes. A 350 will rev over 8,000 rpm nervously while the 302 hums along quite happily provided the valve train can keep up.

Calculating RPM Limits

Suppose you have a 427 big block Chevy with big heavy pistons, a heavy piston pin, and standard ring package. You want to determine a safe rev limit that won't exceed the structural limits of your stock pistons and rods. Just rearrange the formula to solve for RPM.

RPM = (desired piston speed x 6) ÷ stroke

The 427 has a 3.76-inch stroke and we want to impose an arbitrary piston speed of 3,800 fpm.

RPM = (3800 x 6) ÷ 3.76 = 6,063 rpm

To meet your arbitrary piston speed limit, you should set your rev limit to 6,100 rpm. Now, do most people rev 427 Chevy's higher than that? Of course, but if you want to set a limit, that's how you calculate it. Most engines have considerable latitude with modern components. Consider a NASCAR engine for another example. The displacement limit is 358 ci and builders are limited to a maximum bore size of 4.185 inches. We know they run different configurations for different tracks, but let's assume the large 4.185 bore for maximum breathing capability. Using the formula for finding stroke, we calculate:

Stroke = [358 ÷ (4.185² x .7854 x 8)] = 3.25 inches

So what's the mean piston speed of a Cup engine at 9,200 rpm at the end of the back straightaway?

Piston Speed = (3.25 x 9200) ÷ 6 = 4,983 fpm

Of course they have state-of-the-art lightweight components, but now that you know their limits for a 500-mile race, you can adjust your thinking accordingly. And isn't it interesting that a Formula 1 engine and a NASCAR engine both make about the same power. The Formula 1 engine is less than half the size of the NASCAR engine and has to rev twice as fast to do it. Note also that the Formula 1 engine actually runs less mean piston speed than a Sprint Cup NASCAR engine. If you think about these relationships, you can't help but marvel that they are physically possible. And you thought science was boring in high school.

Engine Balancing and Overbalancing

Engine balancing is often thought of as a "black art" practiced by machine shop wizards, but it's not all that mysterious. Thousands of highly competent engine shops do it every day and rarely experience balance-related engine problems. Balancing has become even more precise with today's modern computer-controlled equipment. Balancing components within 2 grams used to be commonplace in performance circles, but not anymore. Most high performance engine builders now balance to within 1/2 gram or less for maximum precision and engine smoothness. So what

Piston Speed Issues

As described in the text, piston speed is an issue for component durability as it relates to engine speed and inertia loading. But there is another factor of near equal importance. You'll note from the discussion that even NASCAR and Formula 1 engines don't like to exceed 5,000 feet per minute mean piston speed. The reason is oil control. In most engines the pistons and rings are lubricated by oil splash being thrown off the spinning crank and rods. NASCAR and Formula 1 engines have an additional source provided by pin oilers which spray a small jet of oil at the bottom of the piston to oil the piston pin and cool the piston crown. But most engines rely only on splash. This becomes a potential problem when you dry sump an engine to remove all oil from the pan. In many cases efforts are also made to isolate the cam tunnel to eliminate cam and lifter oil from dripping on the crankshaft. All of these things combine to reduce the amount of oil being splashed on the cylinder walls. Why is this important?

At elevated engine speeds and the attending high piston speeds, reduced cylinder oiling makes it difficult to maintain a consistent hydrodynamic oil film between the piston and the cylinder wall. When mean piston speed exceeds 5,500 rpm, inconsistent lubrication causes the piston to scuff and, worst case scenario, friction weld itself to the cylinder wall. This typically results in cylinder wall scoring and ultimately engine failure. The greater the piston speed, the bigger the problem, especially when other oiling system modifications combine to reduce cylinder wall oiling. When piston speed is too high, the minimal fog of lubricant in the crankcase has no time to attach itself to the cylinder wall. It's one more reason why Formula 1 engines run such short strokes to reduce piston speed.

does that mean? Everything in modern performance engines is lighter and potentially more fragile, particularly in a high-speed environment. The slightest imbalance has the potential to crack crankshafts, pound out bearings and pistons skirts, and cause other types of mechanical mayhem.

Calculating Balance Weight

The difficulty with engine balance is that some of the parts go round and round while others go up and down. Getting them to do it harmoniously requires precision engine balancing. Adjustable bob weights are used to simulate the weight of the parts during balancing. Rotating weight includes the big end of the rod, rod bolts and rod bearings, plus a small amount (2 to 3 grams) to simulate the oil between the crank journals and bearings. Reciprocating weight includes the small end of the rod, the piston, piston pins, piston rings and retainers if they are used, and a few grams for the oil that clings to the moving parts. Once all of the component weights are equalized, the bob weights are calculated. A normal bob weight comprises 100 percent of the rotating weight and 50 percent of the reciprocating weight. The crankshaft is electronically balanced with the bob weights attached and normal balance is easily achieved.

Overbalancing

High-RPM engines are frequently overbalanced to improve the high-speed balance with less regard to low speed smoothness. The intent is to further smooth the engine's state of balance in its intended operating range. The trick is to balance the assembly so that any critical imbalance falls outside of this range (either above or below it). To accomplish this, the bob weights are adjusted from the calculated norm. Instead of adding 50 percent of the reciprocating weight, the percentage is often increased to something in the 52- to 54-percent range. If any of this is truly a "black art" it may be in actually determining the optimum percentage of overbalance. Many builders know this from experience, but new combinations often require a highly educated guess. The best approach attempts to err to the conservative side, say, 51 to 52 percent. If the engine's ultimate performance and smoothness improves within its primary operating range, they may overbalance it a bit more on the next go round.

Normal Balance = 100% rotating weight plus
50% reciprocating weight

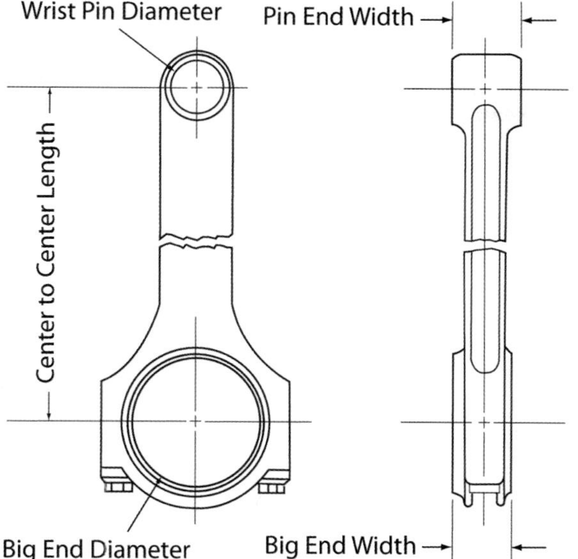

Connecting rod center-to-center length is the primary component in calculating piston position. The rod length is the fixed dimension from the centerline of the big end of the rod to the centerline of the pin end.

Overbalance = 100% rotating weight plus desired percentage of increase in reciprocating weight

Example: 100% rotating weight plus 52.5% reciprocating weight

The overbalance percentage may cause dramatic vibrations outside of the engine's normal operating range, but it is of no concern since you don't run it here for any length of time. Note that overbalancing is a competition engine practice and not something that you would normally do to a street or street/strip engine that has to operate over a broader RPM range. For race engines, it has the potential to save parts and improve performance by reducing vibrations that might be harmful to ring seal, valve train dynamics, and other factors that affect power within a specific power band. It's just one more trick in the high-performance engine builder's bag.

Calculating Piston Position

For the purpose of accurate camshaft selection, it is important to know the piston position where the maximum pressure drop is created in the cylinder. If you recall our core mission of maximizing VE, it's easy to grasp the critical relationship between crank angle and the timing of valve action relative to piston position. For a given rod length, the piston will achieve maximum velocity at some point during the stroke. This point (crank angle) varies according to rod length since stroke length is fixed. You may have heard that a longer rod makes more power, but that's not necessarily the case. As a rule, changes in rod length tend to move the power peak closer to or farther from the torque peak RPM, depending on the change. When rod length is increased, the horsepower and torque peaks move closer together, but the peak values may not change significantly. Shortening the rod tends to separate the peaks farther, which may or may not be beneficial depending on the application. A longer rod causes the piston to linger longer in the vicinity of TDC and the rate of acceleration and deceleration is diminished. To some small degree this provides a little more time for combustion pressure to rise higher before the power stroke. This is beneficial in high-RPM applications where combustion time is limited. In Chapter 14, note that computerized engine simulation programs pay particular attention to rod length versus crank angle. The important thing to remember is the crank angle where the piston achieves maximum velocity and maximum pressure drop within the cylinder. The intake valve opening at this point must be sufficient to maximize flow. Simulations help you calculate and visualize this point.

For those who wish to calculate piston position in the bore relative to crank angle, keep in mind that the point of maximum velocity will differ according to rod length. Generally this point occurs when the rod centerline is 90 degrees to the crank pin, but the actual crank angle will vary with rod length. Use the following formula to calculate piston position relative to crank angle for any given combination of stroke and rod length.

$$P = S\,(1 - \cos C) + (S \times S) \div L\,(\sin^2 \text{ of } C)$$

Where:
S = stroke length
L = rod length
P = piston position relative to deck surface
C = crank angle relative to cylinder centerline
Cos C = cosine of angle C
$\sin^2$ = the sine squared of angle C

This formula is valid for any given value of angle C. While longer rods are generally preferred for high-RPM operation where burn time is minimal, shorter rods have higher acceleration rates and less dwell time around TDC. Higher acceleration rates equate to quicker exposure to the pressure drop in the cylinder, which tends to separate the peaks and promote greater efficiency at lower engine speeds. The difference is often subtle, but savvy engine builders use these tools to accurately position the peaks they want relative to specific applications. Moving the peaks closer together may bring more effective power to bear on a super speedway or a Bonneville engine, while separating the peaks may be more useful for a circle-track or road-racing engine where a broader power band is more desirable. In either case, doing the math often helps illuminate the way.

Piston Speed Formulas at a Glance

Mean Piston Speed$_{fpm}$ = (stroke x rpm) ÷ 6

Mean Piston Speed$_{mph}$ = (mean piston speed$_{fpm}$ x 60) ÷ 5,280

or

Mean Piston Speed$_{mph}$ = (stroke x rpm) ÷ 528

Max Piston Speed$_{fpm}$ = [(stroke x π) ÷ 12] x RPM

Max Piston Speed$_{mph}$ = (max piston speed x 60) ÷ 5,280

Max Piston Acceleration$_{ft/sec^2}$ =
    (rpm^2 x stroke ÷ 2,189) x 1⅓

RPM Limit = (desired piston speed x 6) ÷ stroke

Piston Position = S (1 - Cos C) + (S x S) ÷ L (Sin^2 of C)

Where:

S = stroke length

L = rod length

P = piston position relative to deck surface

C = crank angle relative to cylinder centerline

Cos C = cosine of angle C

Sin^2 = the sine squared of angle C

BRAKE HORSEPOWER AND TORQUE

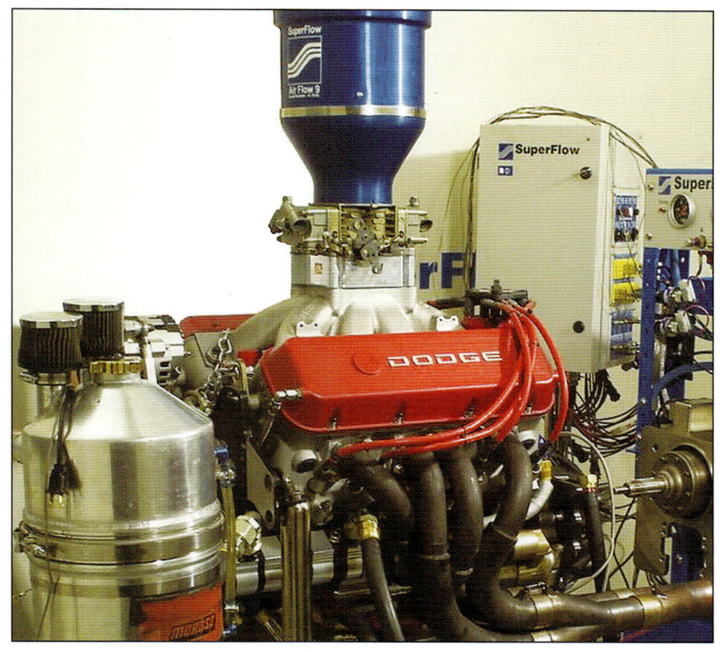

An engine dyno is often called a brake because it brakes or resists engine output for the purpose of measurement. A very early dyno called a prony brake may have also lent its name to the dyno. It used a form of brake shoe to resist engine power. Most modern dynos use a water brake absorber or an eddy current absorber, which uses electrical current to brake the engine.

$$1 \text{ HP} = 33,000 \text{ lb-ft/min}$$

or

$$1 \text{ HP} = 550 \text{ lb-ft/sec}$$

Inventor James Watt first coined the term horsepower in 1780. Accounts vary as to how he came up with it, but the generally accepted version involves trying to develop a method of calculating the amount of work performed by a draft horse operating a pump to remove water from a coal mine. Watt had modified a steam engine to improve its performance and he sought a way to quantify its power by relating it to that of a draft horse. Since horses were the primary power source of that period, the term horsepower was applied to his work. Through observation and measurement, Watt determined the horse's ability to generate a torque (twisting force) about a capstan that operated the mine pump. He calculated that the horse could move 33,000 pounds 1 foot in 1 minute. He called that 1 horsepower.

In engineering speak, work is defined as a force times a distance and it is expressed in ft-lbs while torque is expressed in lbs-ft. Oddly enough, it is commonly referred to as foot pounds and I will follow that convention from here. If you study Watt's calculation you'll see that applying 1 horsepower for 1 minute produces 33,000 lbs-ft of work because a time element is introduced. If you divided 33,000 by 60 (seconds) you get 550 lbs-ft/sec. Watt's calculations were based on the horse generating a force around the circumference of a circle with a lever arm representing the radius much like the throw of a crankshaft. Working backward, you can convert the rotational force of a crankshaft into horsepower by multiplying 2π (two revolutions or 6.2831853) x RPM x torque divided by 33,000. To simplify, divide 33,000 by 2π (6.2831853) to get 5,252, which is the constant applied to the following formulas:

$$HP = (torque \times RPM) \div 5{,}252$$

$$Torque = (HP \times 5{,}252) \div RPM$$

$$RPM = (HP \times 5{,}252) \div torque$$

Note that 5,252 is constant throughout these formulas while RPM, horsepower, and torque are the variables. That's important because 5,252 represents the fixed RPM where torque and horsepower are equal. It is constant. If you're looking at a dyno sheet, the torque and horsepower numbers should match at 5,252 rpm.

And if the dyno curve is presented graphically, the horsepower and torque curves always cross at 5,252 rpm. At that point, torque begins to fall off while horsepower continues to rise. At any point along the graph or chart, you can calculate one value from the other by using the constant 5,252. This one fact makes sure you never get cheated by a fabricated dyno sheet.

All engines generate a particular torque signature based on displacement, engine speed, VE, and flow path dynamics. Not surprisingly, all engines are influenced by specific architecture (i.e., I4, I6, V-6, V-8, V-10, V-12, etc.), each of which applies different attributes to cylinder filling, mean net torque, and overall engine smoothness. Every combination generates a torque peak or "sweet spot" where its particular tuning dynamics achieve maximum volumetric efficiency. In the case of competition engines, this often exceeds 100 percent VE, sometimes by a considerable margin.

The old adage that an engine is an air pump is assuredly true, but we might also think of it as an air processor. Power is governed by the amount of air the engine can process over time and the brake specific fuel consumption (BSFC) generated by the efficiency of the specific component mix. It's relatively easy to supply enough fuel, but it is considerably more difficult to maximize airflow without the aid of a power adder. For any given collection of parts, an engine achieves a torque peak influenced predominantly by intake and exhaust tuning relative to its size or displacement. Through attentive manipulation of these and contributing component hardware, the torque curve can be shaped and positioned to suit the engine's final application. This is a principal focus of all competent engine builders and it begins with the pursuit of VE relative to the engine's static air capacity.

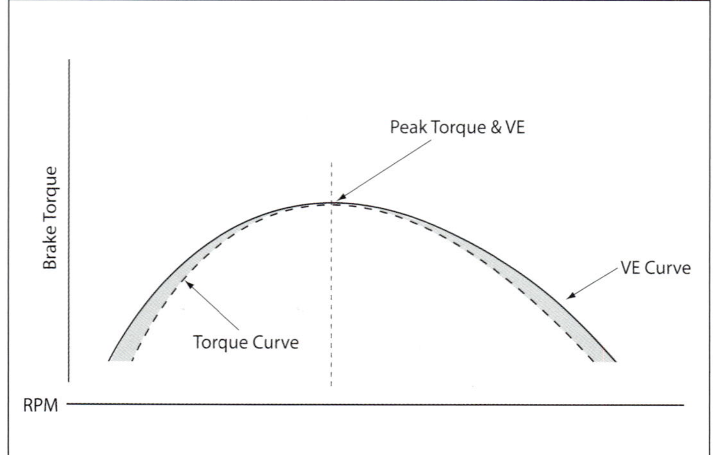

An engine's VE curve mimics the torque curve particularly at the torque peak. As shown here by the shaded areas, it falls off at lower RPM due to poor mixture quality and insufficient inlet airspeed. On the upper end it is limited by insufficient time to fill the cylinder due to increasing RPM.

A load cell attached to the dyno absorber outputs an electrical signal consistent with the amount of torque being absorbed by the dyno. The dyno software converts this signal (voltage) to a torque reading and then calculates horsepower. (Courtesy SuperFlow Technologies Group)

The air mass component depends largely on available air density and the VE a specific component mix is capable of generating. It is primarily governed by inlet and exhaust flow path dynamics, combustion chamber efficiency, valve timing and elements of the bottom end and valvetrain that dictate final RPM capability. As shown in the illustration on page 50, the shape of the torque curve matches the VE curve at peak torque. This is the point of maximum engine efficiency and it typically reflects the lowest wide-open-throttle (WOT) brake specific fuel consumption numbers. Below the peak torque trails the VE curve due to reduced combustion efficiency caused by inadequate intake flow velocity, air/fuel separation, issues and poor mixture quality. Above the torque peak, torque and VE decline due to insufficient time for cylinder filling caused by rising engine speed (RPM).

Fortunately there are methods to address inefficiency on either side of the torque peak and inflate the overall torque curve. This refers to the "area under the curve" and seeks to expand the torque curve in all directions. Horsepower, being a function of torque, follows faithfully. More importantly, a broader torque curve often produces greater acceleration even with a slight reduction in peak torque because it applies more torque over a broader range. If the ideal mix of engine components targets an engine speed range most beneficial to the application, superior performance will accompany it. Complementing these performance gains with appropriately matched gearing and tire combinations ultimately leads to faster cars and better racing all based on the effective production and utilization of torque. This works well even for engines operating well above the torque peak because the upper end of the torque curve expands, thus contributing more horsepower to the car's performance.

Calculating Horsepower from Torque

Now let's calculate some basic torque and horsepower numbers. Again we'll use our 2010 Camaro SS as a basis for our calculations. Its published numbers are as follows:

$$HP = 426 \text{ bhp @ } 5,900 \text{ rpm}$$
$$Torque = 420 \text{ ft-lbs @ } 4,600 \text{ rpm}$$

If the power peak is really 5,900 rpm, how much torque is the engine generating at that engine speed?

$$Torque = (HP \times 5,252) \div RPM$$
$$Torque = (426 \times 5,252) \div 5,900 = 379.2 \text{ ft-lbs}$$

Pretty darn good. Note that it is still making more than 1 ft-lb of torque per cubic inch even at the power peak. That implies an engine that pulls very hard upstairs.

Now, solve for horsepower at the torque peak:

$$HP = (torque \times RPM) \div 5,252$$
$$HP = (420 \times 4,600) \div 5,252 = 367.8 \text{ hp}$$

Even better. Since we're only at 4,600 rpm and the engine is approaching 1 horsepower per cubic inch, we can infer that it will exceed one horsepower per cubic inch by the time it reaches the 5,252-rpm crossover point. We can't know exactly without a full dyno sheet, but lets assume that torque falls off to 400 ft-lbs by 5,200 rpm:

$$HP = (400 \times 5,200) \div 5,252 = 396 \text{ hp}$$

Well above 1 horsepower per cube and still climbing. That's a convenient way of looking at the torque and horsepower relationship based on a contemporary engine with published numbers.

How to Read a Dyno Sheet, Part 1

A properly equipped engine dyno can reveal a wealth of information about an engine besides torque and horsepower. It is really a high-end data-acquisition system that also records all the various pressures, temperatures, vacuum, and about a hundred other things that are measured or calculated. I discuss many of them later in the book so you may want to bookmark this page for future reference. Right now we're going to break down a sample dyno sheet and use it to see how torque and horsepower relate to each other in our calculations. (See dyno sheet on page 52.)

Note that the dyno sheet begins with entries to indicate a test number, a date, time, and operator. Then it asks for a description of the engine being tested and the type of test being performed. There are three basic

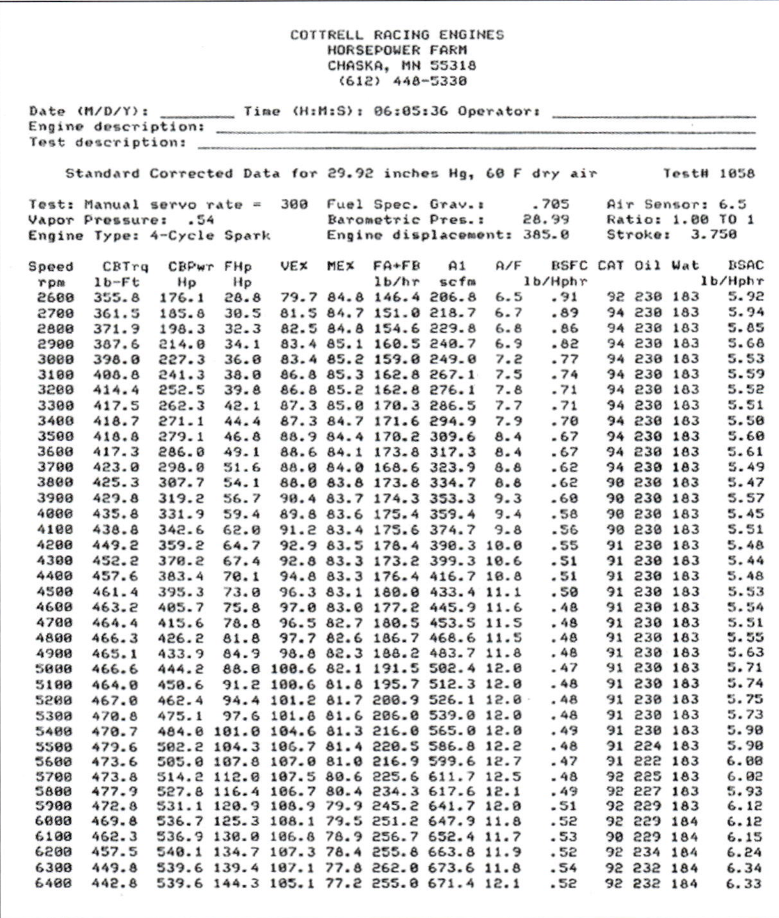

```
                    COTTRELL RACING ENGINES
                       HORSEPOWER FARM
                       CHASKA, MN 55318
                        (612) 448-5330

Date (M/D/Y): _________  Time (H:M:S): 06:05:36 Operator: _____________
Engine description: ___________________________________________________
Test description: _____________________________________________________

     Standard Corrected Data for 29.92 inches Hg, 60 F dry air       Test# 1058

Test: Manual servo rate =  300  Fuel Spec. Grav.:      .705   Air Sensor: 6.5
Vapor Pressure:  .54            Barometric Pres.:     28.99   Ratio: 1.00 TO 1
Engine Type: 4-Cycle Spark     Engine displacement: 385.0    Stroke:  3.750
```

Speed rpm	CBTrq lb-Ft	CBPwr Hp	FHp Hp	VEX	MEX	FA+FB lb/hr	A1 scfm	A/F	BSFC lb/Hphr	CAT	Oil	Wat	BSAC lb/Hphr
2600	355.8	176.1	28.8	79.7	84.8	146.4	206.8	6.5	.91	92	230	183	5.92
2700	361.5	185.8	30.5	81.5	84.7	151.0	218.7	6.7	.89	94	230	183	5.94
2800	371.9	198.3	32.3	82.5	84.8	154.6	229.8	6.8	.86	94	230	183	5.85
2900	387.6	214.0	34.1	83.4	85.1	160.5	240.7	6.9	.82	94	230	183	5.68
3000	398.0	227.3	36.0	83.4	85.2	159.0	249.0	7.2	.77	94	230	183	5.53
3100	408.8	241.3	38.0	86.8	85.3	162.8	267.1	7.5	.74	94	230	183	5.59
3200	414.4	252.5	39.8	86.8	85.2	162.8	276.1	7.8	.71	94	230	183	5.52
3300	417.5	262.3	42.1	87.3	85.0	170.3	286.5	7.7	.71	94	230	183	5.51
3400	418.7	271.1	44.4	87.3	84.7	171.6	294.9	7.9	.70	94	230	183	5.50
3500	418.8	279.1	46.8	88.9	84.4	170.2	309.6	8.4	.67	94	230	183	5.60
3600	417.3	286.0	49.1	88.6	84.1	173.8	317.3	8.4	.67	94	230	183	5.61
3700	423.0	298.0	51.6	88.0	84.0	168.6	323.9	8.8	.62	94	230	183	5.49
3800	425.3	307.7	54.1	88.0	83.8	173.8	334.7	8.8	.62	90	230	183	5.47
3900	429.8	319.2	56.7	90.4	83.7	174.3	353.3	9.3	.60	90	230	183	5.57
4000	435.8	331.9	59.4	89.8	83.6	175.4	359.4	9.4	.58	90	230	183	5.45
4100	438.8	342.6	62.0	91.2	83.4	175.6	374.7	9.8	.56	90	230	183	5.51
4200	449.2	359.2	64.7	92.9	83.5	178.4	390.3	10.0	.55	91	230	183	5.48
4300	452.2	370.2	67.4	92.8	83.3	173.2	399.3	10.6	.51	91	230	183	5.44
4400	457.6	383.4	70.1	94.8	83.3	176.4	416.7	10.8	.51	91	230	183	5.48
4500	461.4	395.3	73.0	96.3	83.1	180.0	433.4	11.1	.50	91	230	183	5.53
4600	463.2	405.7	75.8	97.0	83.0	177.2	445.9	11.6	.48	91	230	183	5.54
4700	464.4	415.6	78.8	96.5	82.7	180.5	453.5	11.5	.48	91	230	183	5.51
4800	466.3	426.2	81.8	97.7	82.6	186.7	468.6	11.5	.48	91	230	183	5.55
4900	465.1	433.9	84.9	98.8	82.3	188.2	483.7	11.8	.48	91	230	183	5.63
5000	466.6	444.2	88.0	100.6	82.1	191.5	502.4	12.0	.47	91	230	183	5.71
5100	464.0	450.6	91.2	100.6	81.8	195.7	512.3	12.0	.48	91	230	183	5.74
5200	467.0	462.4	94.4	101.2	81.7	200.9	526.1	12.0	.48	91	230	183	5.75
5300	470.8	475.1	97.6	101.8	81.6	206.0	539.0	12.0	.48	91	230	183	5.73
5400	470.7	484.0	101.0	104.6	81.3	216.0	565.0	12.0	.49	91	230	183	5.90
5500	479.6	502.2	104.3	106.7	81.4	220.5	586.8	12.2	.48	91	224	183	5.90
5600	473.6	505.0	107.8	107.0	81.0	216.9	599.6	12.7	.47	91	222	183	6.00
5700	473.8	514.2	112.0	107.5	80.6	225.6	611.7	12.5	.48	92	225	183	6.02
5800	477.9	527.8	116.4	106.7	80.4	234.3	617.6	12.1	.49	92	227	183	5.93
5900	472.8	531.1	120.9	108.9	79.9	245.2	641.7	12.0	.51	92	229	183	6.12
6000	469.8	536.7	125.3	108.1	79.5	251.2	647.9	11.8	.52	92	229	184	6.12
6100	462.3	536.9	130.0	106.8	78.9	256.7	652.4	11.7	.53	90	229	184	6.15
6200	457.5	540.1	134.7	107.3	78.4	255.8	663.8	11.9	.52	92	234	184	6.24
6300	449.8	539.6	139.4	107.1	77.8	262.0	673.6	11.8	.54	92	232	184	6.34
6400	442.8	539.6	144.3	105.1	77.2	255.0	671.4	12.1	.52	92	232	184	6.33

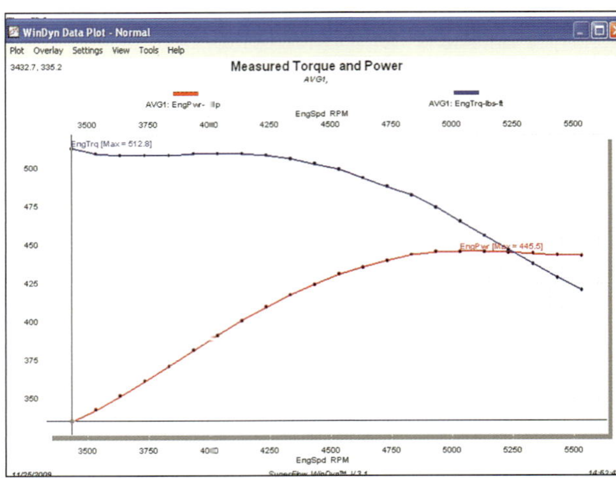

This SuperFlow engine dyno plot illustrates how the torque and horsepower curves cross at 5,252 rpm as described in the text.

The adjacent dyno printout from Cottrell Racing Engines provides all relevant information for one sweep test. See the text to learn how to evaluate and verify the information found in typical printouts such as this one.

test modes: steady state at a selected RPM, a sweep test where the engine is accelerated (unloaded) through a selected RPM range at a fixed rate, and a step test where the engine is held at predetermined RPM intervals until a stable reading is taken at each RPM level.

Step tests are usually run in 250- or 500-rpm increments through a selected RPM range, say, 2,500 rpm through 6,500 rpm for example. The dyno sheet shown here tells us that the test is a sweep or acceleration test at a rate of 300 rpm per second, starting 2,600 rpm and ending at 6,400 rpm (in fact a controlled unloading of the power absorption unit, not an actual acceleration test). The test takes about thirteen seconds at that rate. Note that the test also specifies the specific gravity of the fuel, the barometric and vapor pressures, engine displacement, and stroke. These are entered into the computer prior to the test.

Take a look at the first three columns: RPM, corrected torque (CBTrq), and corrected power (CBPwr). For any given RPM, you can use a variant of the horsepower formula to verify either the torque or the horse-power to see if they are correct. Let's try the torque at 4,000 rpm and see if the horsepower computes properly.

$$HP = TQ \times RPM \div 5,252$$

$$HP = (435.8 \times 4,000) \div 5,252 = 331.91$$

Now try the horsepower at 5,000 rpm and see if the torque is correct.

$$TQ = HP \times 5,252 \div RPM$$
$$TQ = (444.2 \times 5,252) \div 5,000 = 466.587$$

Right on the money with minor rounding. Now examine what's going on at the crossover point of 5,252 rpm. At 5,200 rpm we see 467.0 for torque and 462.4 for horsepower. Why aren't they equal? Because the engine speed component is not exactly 5,252 rpm. Looking at the sheet we might infer that torque at 5,252 rpm is 468.9 ft-lbs by averaging the numbers we see at 5,200 rpm and 5,300 rpm.

$$HP = TQ \times RPM \div 5{,}252$$

$$HP = 468.9 \times 5{,}250 \div 5{,}252 = 468.72 \text{ hp at } 5{,}250 \text{ rpm}$$

Note that we selected 5,250 rpm, not 5,252 rpm. It seemed logical to split the difference between 5,200 and 5,300 rpm, but in fact the numbers are still slightly mismatched. If we actually use 5,252 rpm for the calculation, the answer comes out exact.

$$HP = 468.9 \times 5{,}252 \div 5{,}252 = 468.9 \text{ hp at } 5{,}252 \text{ rpm}$$

The relationship is balanced based on the constant established by Watt in 1780. Later in the book, we'll see how you can relate these torque and horsepower numbers to other recorded data for tuning purposes.

Horsepower and Torque Ratings

It's a good idea to keep dyno numbers in perspective. They are typically referred to as either gross or net figures. Gross power numbers indicate the engine's performance potential under ideal conditions such as within a dyno cell. This frequently means headers with no restrictive exhaust system and no parasitic losses from auxiliary components such as a fan, alternator, power steering pump, AC compressor, mechanical fuel pump, and even the water pump in many cases. It is maximum observed power at the flywheel, which is typically corrected to SAE standard J607 or standard temperature and pressure (60-degree dry air and 29.92 Hg barometric pressure). This correction is used by most dyno shops and performance magazine testers.

Or it is corrected to SAE standard J1349 (77-degree dry air and 29.93 Hg) as used by OEM automakers. The difference is about 4 percent; so for comparison testing you should always compare numbers based on the same correction factor. Different correction factors can be mathematically converted, but it is always easier to compare apples to apples.

Net torque and horsepower represent the real world with all the ugly parasitic components in play, including the air cleaner, a full exhaust system, and all the items required to make a vehicle fully functional. Few of these items are ever in place on an engine dyno, but they are there for a chassis dyno test and the numbers vary accordingly. The chassis dyno also accounts for all the friction and inertia losses in the drivetrain and the tires.

We pay closer attention to this today, but in the 1960s and 1970s, they looked at things a little differently. For racing and insurance purposes, the OEMs often underrated brake horsepower (bhp). Typically they published a brake horsepower figure at a specified RPM, but neglected to mention that power kept rising above that point. It was an arbitrary rating, and not necessarily the maximum output.

In 1971 General Motors switched to net ratings, although they still published both ratings for just that one year. All of the other automakers followed them shortly. Today we see only net ratings—isn't it interesting how high they have climbed as automakers improve efficiency? Still, muscle car enthusiasts often long for a way to compare net and gross ratings for their cars. There is no direct conversion factor, but we might infer some educated guesses based on horsepower per cubic inch.

One horsepower per cubic inch used to be the magic number. Among others, the fuel-injected 283-ci small-block Chevy V-8 achieved it back in 1957. Most cars had far less. At 128 gross horsepower, a 263-ci 1952 Buick straight-8 provided 0.48 hp/ci. A 375-hp 396-ci Chevelle offered 0.94 hp/cl in 1966 while the average vehicle was still mired somewhere between 0.5 and 0.8 hp/ci. The 1970–1971 Dodge Challenger R/T had a 390-hp 440 Six-Pack that delivered 0.88 hp/ci.

How about the 1969 Z28 Camaro? Its 290-hp 302-ci engine delivered 0.96 hp/ci; a gross rating that we know was better. The engine made more horsepower above the published RPM and the cars were fast for their weight and displacement. Big-valve iron cylinder heads of the day barely flowed 200 cfm, yet Traco Engineering and GM dyno sheets show more than 400 hp from blueprinted Penske Trans Am engines. That's more than 1.3hp/ci in gross or race trim.

Of course these were race-prepped engines, but if we assume 1.1 hp/ci (relatively easy to achieve today) it is easy to see that the production 302 probably made 330 to 350 hp gross, which may have nettled about 275 in the car. An educated guess, but probably still enough to provide the ETs and speed those cars typically ran. One can only speculate.

Low-compression base engines suffered miserably. The base 1971 Chevy 350 was rated at 245 gross hp, but only

165 net hp. That's 0.47 hp/ci, which is worse than the 1952 Buick straight-8. Go figure. It serves to illustrate how much engines are affected by lower compression ratios, unfavorable fuel and spark curves and excessive parasitic losses in the engine and in the driveline.

Indicated Horsepower

While the engine dyno is an effective tool for measuring an engine's output, it can't directly measure losses within the cylinders due to the friction and inertia of the operating parts. It only measures output at the flywheel, not what is left on the table due to physics. We know that the engine makes torque when combustion pressure forces the piston down and turns the crankshaft. The force applied to the piston top is simply the combustion pressure (cylinder pressure) times the area of the piston top. Since gasoline engines operate on the expansion cycle, cylinder pressure varies greatly during a power stroke. It is greatest just after ignition and falls off quickly as the energy is transferred to the piston to turn the crankshaft.

Spark advance gives the combustion process a head start before the piston reaches TDC, but peak cylinder pressure generally occurs about 12 to 15 degrees after TDC when the piston is just starting to travel down the bore. If we know the cylinder pressure we can calculate the indicated horsepower using a widely accepted formula called PLAN.

$$\text{Horsepower} = (P \times L \times A \times N) \div 33{,}000$$

Where:
P = cylinder pressure (MEP) in pounds per square inch (psi)
L = length of the stroke in feet (stroke ÷ 12)
A = piston area in square inches (bore^2 x 0.7854)
N = number of power strokes per minute [(RPM ÷ 2) x number of cylinders]

And, of course, 33,000 represents 1 hp.
This yields the following equation:

$$\text{Horsepower} = (\text{MEP} \times \text{stroke} \times \text{bore}^2 \times 0.7854 \times$$
$$\text{number of cylinders}) \div (12 \times 2 \times 33{,}000)$$

On closer inspection we find the displacement formula in the middle of our equation. If we already know the displacement we can substitute it to simplify the equation.

$$\text{Horsepower} = (\text{MEP} \times \text{displacement} \times \text{RPM}) \div 792{,}000$$

It's all about cylinder pressure, piston area and leverage applied over time. Not surprisingly, all of these things can be modified to improve an engine's performance. Note that most engine modifications affect cylinder pressure, displacement, and/or engine speed:

- If you shorten the stoke to increase RPM, you increase N, which gives you more power strokes per minute.
- If you overbore the engine, you increase A for more piston area to be acted upon by cylinder pressure.
- If you raise the compression ratio, you increase cylinder pressure (P).

Because MEP only occurs within the cylinder, it is difficult to measure directly. Engine labs like Hi-Techniques insert a small pressure transducer directly into the combustion chamber to measure the mean effective pressure of an engine running on the dyno. (SuperFlow Technologies calls this Engine Cycle Analysis.) A rotary crankshaft encoder is teamed with the pressure transducer to pinpoint mean effective pressure at every degree of crankshaft rotation. The pressure transducer is called an "indicator"; hence, the term Indicated Mean Effective Pressure (IMEP). This mean effective pressure measured on the dyno is the pressure (P) used in the PLAN formula.

For an example, lets take a 360-ci Dodge engine on which the dyno has measured an IMEP of 180 at 4,800 rpm:

$$\text{HP} = (180 \times 360 \times 4{,}800) \div 792{,}000 = 392.7 \text{ hp}$$

The operative word here is "mean." The pressures we are discussing are mean or "average" pressures that occur over the duration of the power stroke. This makes them sound too low, but cylinder pressure varies greatly throughout the power cycle. Just after ignition, the pressure in a high-performance engine may exceed 1,200 psi, but it decays rapidly during the expansion process as it pushes the piston down the bore.

Hi-Techniques rotary encoder system attached to the front of the crankshaft provides the crank position signal that helps determine cylinder pressure relative to piston position and related valve action. Sharp eyes will note that this particular setup is installed on a dry-sump-equipped small-block Chevy race engine mounted on a SuperFlow 901 engine dyno. (Courtesy Hi-Techniques)

If you have a 4.125-inch piston with 13.36 square inches of piston area, that 1,200 psi very briefly becomes more than 16,000 pounds of force pushing on the piston. Midway down the bore, the pressure may be less than half the initial pressure and it drops off to near atmospheric when the exhaust valve cracks open.

Keep in mind that IMEP is an average pressure derived from very high pressure at the ignition point and there is rapid decay from there on down the bore. This average pressure times the leverage of the stroke length and many repetitions (rpm) yield average torque and, thus, the average horsepower that makes your car go fast.

Indicated Torque

Dyno operators tell you that peak MEP occurs at the same RPM as peak torque. That's the point of maximum efficiency, but we can calculate the indicated torque at any RPM provided that we know the IMEP at that point. To calculate indicated torque, multiply the MEP times the exact calculated displacement and divide by 150.8.

$$Torque = (MEP \times displacement) \div 150.8$$

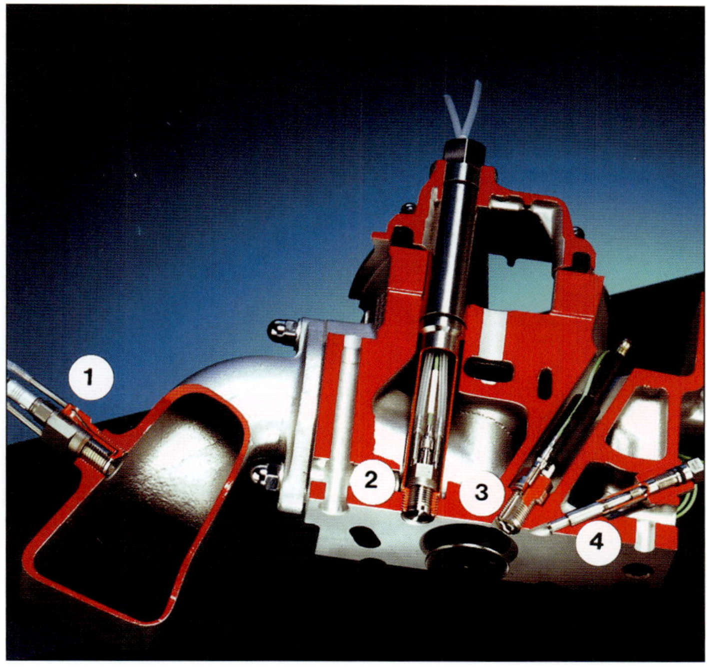

Combustion chamber pressure transducers work in union with the rotary encoder. They directly sample cylinder pressure continuously so the software can relate combustion pressure to crank angle and piston position. This is the indicator that measures mean effective pressure. Several types are shown: 1. intake/exhaust pressure sensor, 2. water-cooled reference-grade pressure sensor, 3. Spark-plug-mounted pressure sensor and 4. an un-cooled pressure sensor. (Courtesy Hi-Techniques and Kistler Corporation)

Revisiting our 360 Dodge engine once again, we simply plug in the variables. The 150.8 is a constant.

$$Torque = (180 \times 360) \div 150.8 = 429.7 \text{ ft-lbs}$$

Now recall that horsepower equals torque times RPM divided by 5,252 and see what you get.

$$HP = 429.7 \times 4800) \div 5,252 = 392.7 \text{ hp}$$

The same answer we get from the MEP calculation.

Brake Mean Effective Pressure

If you already know the measured brake horsepower or brake torque, you can calculate the MEP required to produce it. Simply rearrange the formula as follows:

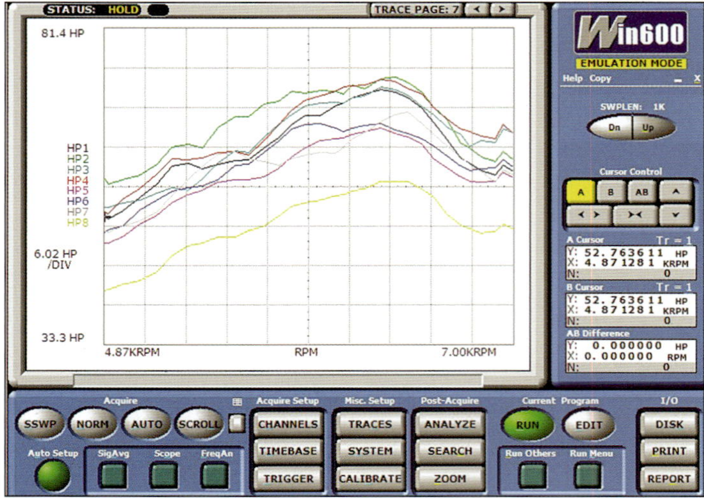

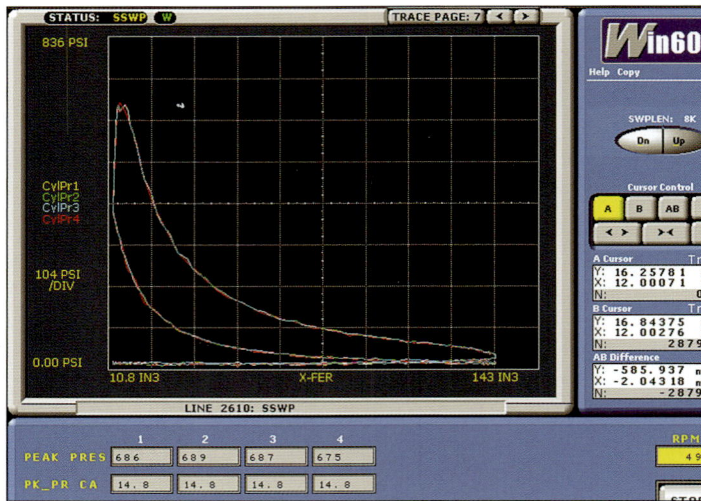

One of the results of precision in-cylinder pressure measurement is this output graph from Hi-Techniques Win600e analysis software. The graph displays the horsepower output of eight individual cylinders from 4,870 rpm to 7000 rpm. Of particular note is the yellow trace (bottom), indicating that cylinder number-8 is not participating equally in the power process.

This screen illustrates an actual pressure volume trace of a four cylinder engine that exhibits a peak pressure anomaly on cylinder number-4, while confirming that net mean effective pressure was not seriously impacted. Nonetheless, sharp engine builders will try to pinpoint the cause of the peak pressure drop in that cylinder. (Courtesy Hi-Techniques)

$$MEP = (horsepower \times 792{,}000) \div (displacement \times RPM)$$

Suppose you have an engine making 500 hp at 6,800 rpm. If the displacement is 383 ci, what is the MEP?

$$MEP = (500 \times 792{,}000) \div (383 \times 6{,}800) = 152.05 \text{ psi}$$

We know from the torque formula that the same 383 makes 386.17 ft-lbs of torque at the same RPM. So we can also calculate the MEP from torque using that figure.

$$MEP = (torque \times 150.8) \div displacement$$

$$MEP = (386.17 \times 150.8) \div 383 = 152.05 \text{ psi}$$

Mean Effective Pressures

To review briefly, the Indicated Mean Effective Pressure (IMEP) is the force acting against the piston top. It is a measured number. Friction Mean Effective Pressure (FMEP) represents the frictional losses between the pistons and cylinders walls and the crankshaft bearings. The difference is the actual output at the flywheel. It is calculated from observed torque on the dyno. It can also be calculated from observed horsepower which is derived from torque.

Combustion pressure times piston area and the leverage imparted to the crankshaft makes our cars go fast. Dyno operators always tune for max torque because horsepower is a function of torque. You might say that torque is the force that gets a car moving and horsepower is what keeps it accelerating. That's pretty close to correct, but the real key to maximum performance is the area under the horsepower curve.

The car with the greatest average horsepower across its entire RPM range will be the faster car even if its opponent has higher peak numbers. The modifications we make are all aimed at increasing the mean effective pressure on the piston. MEPs range from 170 to 185 in most high-performance applications and most racing engines operate at slightly above 200 psi.

Mechanical Efficiency

An engine's mechanical efficiency (ME) is its ability to overcome the frictional losses generated by its moving parts. The question is often asked; why does an engine idle? Well, because at minimum throttle angle it makes just enough torque to overcome its frictional and pumping losses. If we're lucky enough to have both brake and indicated output figures, we can calculate an engine's mechanical efficiency using the following formula.

$$ME = (\text{brake output} \div \text{indicated output}) \times 100$$

Assume an engine that has 488 indicated horsepower at 5,900 rpm and 460 ft-lbs of torque at 4,600 rpm. Its measured brake output is 426 hp at 5,900 rpm and 420 ft-lbs of torque at 4,600 rpm. Calculate its mechanical efficiency from horsepower and then from torque.

$$ME = (426 \div 488) \times 100 = 87.29\%$$

$$ME = (420 \div 460) \times 100 = 91.30\%$$

The difference in mechanical efficiency represents the friction losses within the engine. In this case friction withholds 62 hp and 40 ft-lbs of torque. Note the higher efficiency at peak torque. The engine achieves its best volumetric efficiency at that point and is better able to overcome its parasitic losses.

So why does this matter? Do you suppose those numbers bear a strong resemblance to our 2010 Camaro SS brake output figures? If they do, it means the Camaro engine really has the potential to make 1.3 hp/ci in street trim. You have to have more potential than what the brake shows you in order to overcome frictional losses.

Are these numbers a guess? Sure they are, but think about it. Who is better at finding engine efficiency than the OEM automakers? They want all the power and efficiency they can get so they do everything possible to reduce friction and pumping losses. Racers do the same thing. They use thinner piston rings, narrower bearings, piston and bearing coatings and a whole bag of tricks to increasing MEP So is the Camaro engine really a tamed down racing engine? If you thought that, you might be close to correct.

Brainstorming with MEP

Since mean effective pressure is the master component of torque and horsepower, we can use projected MEP in the torque and horsepower formulas to model displacement and RPM combinations that might suit our particular needs. We also know that most race engines operate at or above 200-psi MEP and that number is influenced by piston

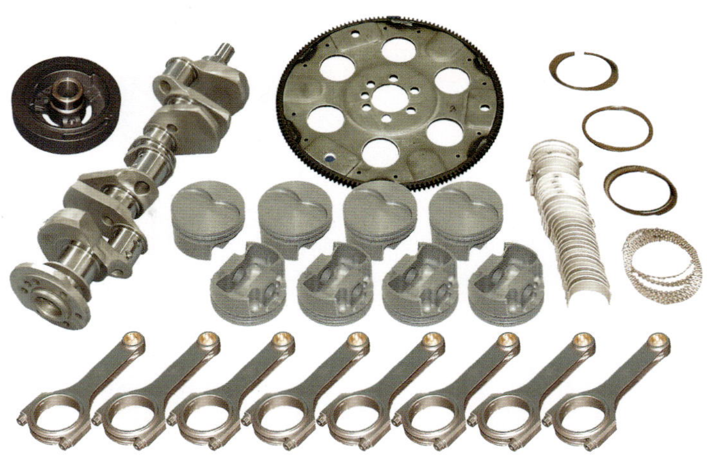

The crankshaft, rods and pistons, rings and bearings, flywheel and harmonic damper all contribute to friction and inertia losses within the engine. (Courtesy Eagle Specialty Products)

area, stroke length and RPM. It is also dependent on how well the cylinder heads and camshaft fill the cylinders. Here's an example:

Say you're trying to break an existing speed record at Bonneville. You've already got drag and frontal area numbers for your car and you have calculated that you need 800 hp to get the job done. Further assume that you are class limited to 372 ci. Of course, traction, aerodynamics, and weather will influence your efforts, but 800 hp seems capable of producing the desired result. Well, that's 2.15 hp/ci, which is right up there with some of the best normally aspirated engines. If you run 14:1 or more compression ratio with good cylinder heads and camshaft timing, you should be able to generate an MEP of 200 psi and maybe even 210 psi. Conservatively assuming that you achieve an MEP of 200 psi, what RPM will produce the desired horsepower given your stated displacement limit?

$$HP = (MEP \times \text{displacement} \times RPM) \div 792,000$$

$$HP = (200 \times 372 \times 7{,}500) \div 792{,}000 = 705 \text{ hp}$$

That's not going to cut it. You're going to have to spin the engine a good bit faster or modify your package to raise the MEP. Let's try more RPM first since it might be easier to achieve. Plug in 8,000 rpm and you only reach 752 hp, so you still need more engine speed. Try 8,500 rpm.

Muscle Cars and MEP

The following chart is a representative list of muscle car engines from the 1950s and 1960s. It shows the calculated mean effective pressure at the both the torque peak and the power peak for each engine. Try using the MEP formulas for torque and horsepower to see if you can calculate the same results. Then, compare the horsepower per cubic inch versus the compression ratio, MEP, and induction to see how these engines stack up against each other. Note that these are all pre-1971 gross power and torque figures.

Car	CID	HP @ RPM	MEP	HP/CI	TORQUE @ RPM	MEP	C/R	Carb
1957 Chevy	283	283 @ 6,200	128	1.0	290 @ 4,400	154	10.5:1	FI
1964 Corvette	327	375 @ 6,200	146	1.15	350 @ 4,400	161	11:1	FI
1966 Chevy Nova	327	350 @ 6,000	141	1.07	360 @ 3,200	166	11:1	4V
1967 Corvette	427	435 @ 5,800	137	1.02	460 @ 4,000	162	11:1	3/2V
1969 Camaro Z/28	302	290 @ 5,800	13	1.96	290 @ 4,200	144	11:1	4V
1969 L88 Corvette	427	560 @ 6,400	162	1.31	n/a			
1970 Camaro Z/28	350	360 @ 6,000	135	1.03	380 @ 4,000	163	11:1	4V
1971 Corvette LS6	454	425 @ 5,600	132	0.93	475 @ 4,000	158	11:1	4V
1970 Buick GSX	455	360 @ 4,600	136	0.79	510 @ 2,800	169	10:1	4V
1957 Cadillac Eldorado	365	325 @ 4,800	147	0.89	400 @ 3,200	165	10:1	2/4V
1951 Chrysler	331	180 @ 4,000	107	0.53	312 @ 2,000	142	7.5:1	4V
1957 Chevy 300D	392	390 @ 5,200	151	0.99	435 @ 3,600	167	10:1	FI
1965 Dodge Race Hemi	426	425 @ 5,600	141	0.99	480 @ 4,600	170	12:1	2/4V
1971 Street Hemi	426	425 @ 5,000	158	0.99	490 @ 4,000	173	10.2:1	2/4V
1957 Ford	312	270 @ 4,500	152	0.86	323 @ 2,300	160	9.7:1	2/4V
1965 SOHC Ford	427	657 @ 7,500	162	1.53	n/a			
1965 T-Bird	427	425 @ 6,000	131	0.99	480 @ 3,700	169	11.5:1	2/4V
1969 Boss Mustang	302	290 @ 5,800	131	0.96	290 @ 4,300	145	10.5:1	4V
1969 Boss Mustang	429	375 @ 5,600	123	0.87	450 @ 3,400	158	11.3:1	4/V
1957 Continental	368	300 @ 4,800	143	0.81	415 @ 3,000	170	10:1	4V
1964 Mercury Marauder	427	425 @ 6,000	131	0.99	480 @ 3,700	169	11.5:1	2/4V
1958 Oldsmobile J2	370	312 @ 4,600	145	0.84	415 @ 2,800	169	10:1	3/2V
1970 Oldsmobile	455	370 @ 5,200	123	0.81	500 @ 3,600	166	10:5	4V
1956 Chrysler 300B	354	355 @ 5,200	153	1.00	405 @ 3,400	162	10:1	2/4V
1964 Max Wedge	426	425 @ 5,600	141	0.99	n/a		13.5:1	2/4V
1957 Pontiac	347	315 @ 4,800	149	0.90	n/a		10:1	FI
1964 GTO	389	348 @ 4,900	144	0.89	428 @ 3,600	166	10.75:1	3/2V
1965 GTO	389	360 @ 5,200	141	0.92	424 @ 3,600	164	10.75:1	3/2V
1965 Pontiac	421	376 @ 5,000	141	0.89	461 @ 3,600	165	10.75:1	3/2V
1970 Ram Air IV	400	370 @ 5,500	133	0.90	245 @ 3,900	168	10.5:1	4V

HP = (200 x 372 x 8,500) ÷ 792,000 = 798 hp

Now that certainly seems possible. If you've built a big-bore small-block with a relatively short stroke it will probably handle up to 8,500 rpm without distress on the dragstrip or even Bonneville's 5-mile dyno. So the answer is a definitive yes. You can get there within achievable RPM limits if you concentrate on building maximum MEP. That means higher compression, cylinder heads with superior breathing, a light, stable ring package for optimum cylinder sealing, and all possible efforts to minimize parasitic losses in the short block. All things that we strive for in a competition engine.

You can see how the horsepower formula can help you predict engine performance based on projected mean effective pressure. There's more to it of course, but if you sweat the details and generate the required MEP within your displacement and RPM limits, you're well on your way to making big power.

Torque & HP Formulas at a Glance

HP = (torque x RPM) ÷ 5,252

Torque = (HP x 5,252) ÷ RPM

RPM = (HP x 5,252) ÷ torque

Horsepower = (P x L x A x N) ÷ 33,000

Horsepower = (MEP x displacement x RPM) ÷ 792,000

Torque = (MEP x displacement) ÷ 150.8

MEP = (horsepower x 792,000) ÷ (displacement x RPM)

MEP = (torque x 150.8) ÷ displacement

Mechanical Efficiency =
(brake output ÷ indicated output) x 100

Friction output = indicated output − brake output

1 hp = 33,000 lb-ft/min

1 hp = 550 lb-ft/se

INDUCTION MATH

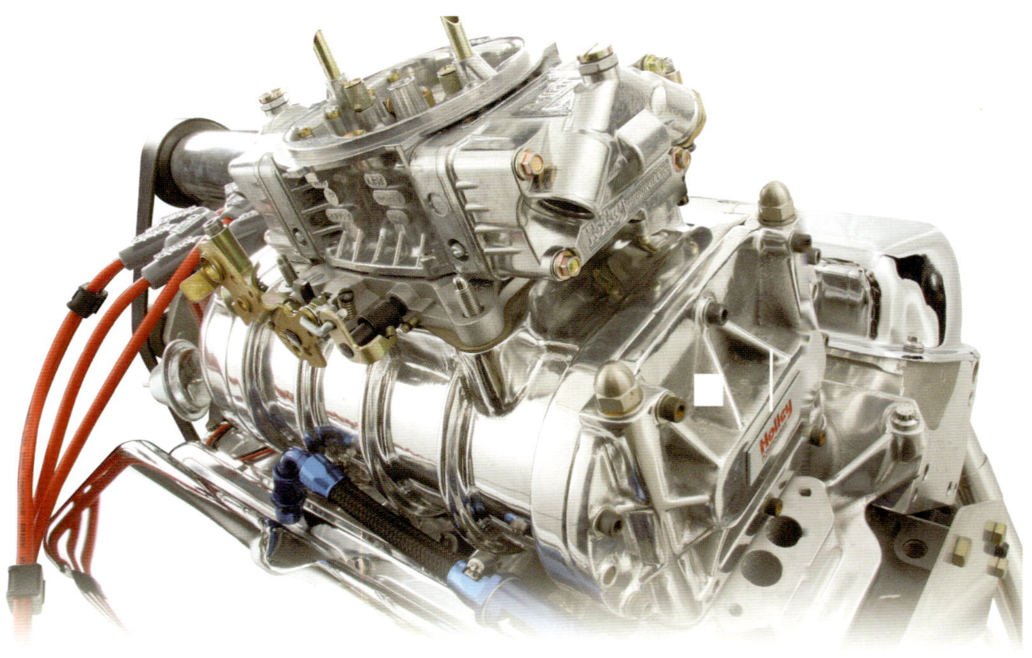

In Chapter 5, I discussed mean effective cylinder pressure and ways to measure or calculate it. When we consider the importance of the relationship between MEP and engine performance, we have to acknowledge where that pressure came from. It comes from the combustion (reaction) of the air and fuel mixture pushed into the cylinder on the intake stroke. The compression stroke squeezes the daylights out of it to increase its density and the spark plug lights it off to create combustion pressure (MEP) during the power stroke. Among other things, the pre-combustion density of the air/fuel charge in the cylinder determines MEP. The energy volatility of that combustion pressure is governed by mixture quality, chamber turbulence, spark timing, and a host of other contributors, but charge density is the compelling factor.

Engine Air Capacity

We can't make a cylinder hold more than its physical dimensions, but we can increase the density of the gas captured within it. Hence the terms air capacity, volumetric efficiency, and inertial ram tuning. The earth's atmosphere does us the courtesy of providing 14.7 psi of pressure to fill the cylinders for free. In a perfect world, the engine would inhale the exact swept volume of all its cylinders every two revolutions of the crankshaft. That represents its theoretical or potential air capacity. The following formula is used to calculate it.

$$\text{Air Capacity}_{cfm} = (\text{displacement} \times \text{RPM}) \div (1{,}728 \times 2)$$

Note that air capacity is a function of displacement (volumetric capacity) and engine speed (RPM). The displacement is divided by 1,728 (the number of cubic inches in a cubic foot) to convert it to cubic feet. The RPM is divided by 2 because the engine only intakes on every other revolution. The formula can be simplified as follows:

$$\text{Air Capacity}_{cfm} = (\text{displacement} \times \text{RPM}) \div 3{,}456$$

In practice this formula is often useful to calculate the engine air requirement at both peak torque and peak power. Why? Because we want to know the minimum requirement at peak torque or the point of highest efficiency. Then we want to establish the maximum requirement at peak power so we can determine the effective airflow range of the engine under wide open throttle (WOT) conditions.

Let's calculate the airflow requirement for a 350-ci engine with a torque peak at 5,300 rpm and a maximum engine speed of 6,500 rpm. First the torque peak:

$$\text{CFM} = (350 \times 5{,}300) \div 3{,}456 = 536 \text{ cfm at the torque peak}$$

And then the power peak:

$$\text{CFM} = (350 \times 6{,}500) \div 3{,}456 = 658 \text{ cfm at the power peak}$$

These are the engine's theoretical air requirements if we didn't have all sorts of interference and restriction from carburetor venturis and throttle plates, manifold runners, intake ports, valves, and so on.

Volumetric Efficiency

In the preceding example the theoretical airflow capacity requires at least 536 cfm at the torque peak and 658 cfm at the maximum engine speed. These would be airflow requirements at WOT. So we need a 650-cfm carburetor, right? Not so fast. Many factors combine to reduce or in some cases increase the actual airflow requirement.

Volumetric efficiency is the volume of air the engine is theoretically capable of ingesting (potential) versus the actual volume that makes it into the cylinder, all other factors considered. In effect, it is the difference between the mass of the charge consumed by the cylinders and the mass of an equal volume at atmospheric pressure at any given RPM. It is affected by carburetor or throttle body size restrictions, engine speed, air temperature, manifold and port restrictions, valve size, chamber shrouding, camshaft overlap, and pumping losses.

Airflow measured on a running engine divided by the potential or theoretical air capacity at any given RPM is the engine's actual volumetric efficiency at that particular engine speed. While not perfectly linear, it is sufficient for calculating volumetric efficiency. The following formula yields the percentage of volumetric efficiency:

$$\text{VE \%} = (\text{measured CFM} \div \text{potential CFM}) \times 100$$

If our 350-ci sample engine is a street engine and we are able to measure airflow on a dyno we might find that the actual airflow at 5,300 rpm is only 450 cfm and rising to 539 cfm at the maximum engine speed of 6,500 rpm. This would yield volumetric efficiency calculated as follows:

$$\text{VE \%} = (450 \div 536) \times 100 = 84\% \text{ at the torque peak}$$

$$\text{VE \%} = (539 \div 658) \times 100 = 81.9\% \text{ at maximum engine speed}$$

VE is an indicator of how well the induction system does its job of filling the cylinders. Millions of dyno pulls and extensive research have established typical VE percentages for most engines. Standard passenger cars are generally 70- to 80-percent efficient, while high performance engines range from the low 80s to the mid 90s. Tuned racing engines routinely exceed 100-percent volumetric efficiency because their high-flow cylinder heads, tuned intake systems, cam timing, and exhaust scavenging provide exceptional efficiency. Highly refined Pro Stock drag racing engines are capable of achieving up to 125-percent VE. That means that the actual charge in the cylinder is 25 percent denser than the same volume before it enters the engine. Combined with ultra-high compression ratios, it really packs some power in the form of mean effective pressure.

At 85-percent VE, a 350-ci street engine only consumes 297.5 ci of air by volume. The only way to change it is to supercharge it or increase the efficiency of cylinder

---

Pro Stock and Pro Mod drag racing engines achieve VE percentages as high as 125 percent due to their narrow power band and the exceptional high speed efficiency of tunnel ram intake manifolds and split-Dominator style carburetors designed to take advantage of induction system wave tuning. (Courtesy Don Cooper/Reher & Morrison)

filling by applying the right combination of naturally aspirated components. In a naturally aspirated engine the goal is to increase the inertia of the incoming charge and reduce the restrictions it has to navigate on its way to the cylinder. A good system accomplishes this with a mean port speed of about 240 ft/sec. The important thing to remember is that low restriction and high charge inertia are achievable through proper carburetor or throttle body sizing, appropriate manifolding and cylinder head port volumes that match engine displacement and anticipated RPM.

Power and Volumetric Efficiency

The cornerstone of power building is volumetric efficiency (VE). The more air an engine is able to process, the greater its power potential. Volumetric efficiency is determined according to an engine's static air capacity or displacement. A displacement of 400 ci represents 100 percent air capacity for an engine of that particular size. At any given engine speed a percentage of that volume is being processed into torque, depending on a host of variables that conspire to limit airflow. Without these pesky restrictions, atmospheric pressure can easily fill the cylinders 100 percent every two crankshaft revolutions. In practice, this is difficult to achieve because airflow is restricted by a throttling device (carburetor, throttle body, or other), imperfect intake manifolding, intake ports, valves, and all the attending flow restrictions and pressure dynamics present in a running engine. Hence, VE in a production engine rarely exceeds 70 to 80 percent.

As previously noted, VE is reduced below the torque peak due mainly to insufficient airflow and poor mixture quality. Above the torque peak, VE is limited by inadequate time to fill the cylinder due to RPM. One of the successful engine builder's primary goals is to exceed the static air capacity of the engine and optimize combustion efficiency once fuel is introduced to the process. Savvy engine builders skillfully manipulate the component composition to accomplish this—broadening the torque curve and positioning it to best suit the intended application.

In specifying components to meet VE requirements, builders target intake ports, dimensional qualities of intake manifolds and exhaust headers, carburetor size, rod-to-stroke ratios, valve timing, and static compression ratio. The specific component matrix is adjusted to suit the application's operational requirements. Oval-track and road-racing engines typically call for a component mix producing a broad torque curve over a wide range of RPM. This affords the engine builder an opportunity to tune the intake and exhaust systems separately to effectively broaden the power band. Conversely, drag racing applications seek a higher and narrower power band in which intake and exhaust tuning are more closely aligned.

Identifying and targeting the required power band is one of the engine builder's first steps. Since VE and engine speed are closely aligned, it is critical to target VE modifications to the desired engine speed. If a drag racing engine leaves the starting line at 7,000 rpm and cycles

VE and Trapped Air Mass

Volumetric efficiency is how we rate an engine's ability to process air. Most VE numbers, even those generated on a dyno, are slightly misleading because they don't accommodate all the details of the cylinder-filling and combustion process. On a dyno, the air is measured going into the engine and VE is calculated according to displacement, but there are other considerations. Since power can only be generated from the actual "trapped mass" of the air and fuel mixture in the cylinder at the time of ignition, actual volumetric efficiency must also consider losses that occur during the camshaft overlap period and the amount of blowby (cylinder pressure) that escapes past the rings. Hence, the trapped mass is actually the measured CFM minus the CFM loss from overlap and blowby divided by the theoretical CFM.

Trapped mass = measured CFM − (blowby CFM + overlap CFM) ÷ theoretical CFM

Without proper equipment, blowby CFM and overlap CFM are very difficult to measure. Engine labs use calibrated blowby meters to actually measure the pressure differential in the crankcase and calculate the amount of loss through the rings while sophisticated higher math is employed to determine the amount lost when both valves are open simultaneously during the overlap period. In a sense you can think of blowby and overlap CFM as leaks that permit potential power to escape without delivering the full amount of power that is theoretically available (see arrows).

Total trapped mass in the combustion space is subject to loss via leakage from blowby past the rings and reversion into the intake tract during the overlap period (see arrows).

between there and 9,000 rpm through the gears, its VE at 5,000 rpm is largely irrelevant. And, of course, an engine delivering power between 4,500 rpm and 7,200 rpm will need broader tuning efficiency from its parts combination. Hence, airflow management within the targeted engine speed range becomes a central challenge in matching or exceeding an engine's potential VE capacity.

Intake Manifolds

Intake manifold and carburetor restrictions are prevalent in many types of racing. They are primarily intended to limit airflow and RPM potential. In some cases more than one choice is offered and the final selection is based on which configuration generates the best VE and torque tuning potential. That's why, where rules permit, a twin-carb high-RPM tunnel ram is chosen over a single 4-barrel for a drag racing application, but you're not likely to see a tunnel-ram intake on a road racing car. Some classes dictate the use of a dual-plane intake, which often extends to the use of a stock cast-iron manifold. When the intake manifold is specified, all you can do is identify the manifold's characteristics and tailor your package accordingly.

We'll also discuss how to map manifold characteristics on a flow bench to obtain a ballpark view of individual port strengths and weaknesses. Once you have a clear picture of the manifold's efficiency you can evaluate potential steps to ensure its contribution to maximum performance. Depending on other restrictions, these may include rocker ratio or cam timing adjustments to individual cylinders based on individual runner flow dynamics. Or it may be addressed by manipulation of header dimensions to complement and possibly broaden the torque range dictated by the intake manifold's fixed dimensions. If allowed, carb spacers may support better mixture quality and, in the case of dual-plane intakes, staggering jetting from side to side may also provide some improvement particularly as it relates to the lean side of the engine. Many circle track classes also require a 2-barrel carburetor of a specified size with no modifications allowed, although repositioning of the carburetor location on the intake manifold is sometimes permitted.

Street Carburetor vs Race Carburetor

Engine air capacity and VE are important to carburetor selection because most carburetor functions are initiated and controlled by airspeed through the venturis and boosters. Once you establish the engine's airflow capacity using the CFM formula, you need to multiply it by the appropriate percentage to account for VE losses.

Street Carburetors

The general rule for performance street engines is 85 percent of the theoretical or potential air capacity.

$$\text{Carb Size}_{cfm} = [(\text{displacement x RPM}) \div 3{,}456] \times 0.85$$

Recalling our theoretical 350-ci engine, we can calculate the cfm requirement at maximum engine speed:

$$CFM = [(350 \times 6{,}500) \div 3456] \times 0.85 = 559.5$$

The airflow demand at peak power is only 560 cfm, so a relatively small carburetor suits it well. Depending on the venturi and throttle plate size, a 550-cfm carburetor provides surprisingly crisp throttle response on the street. You are in the ballpark with a 600-cfm carb as well and, since most performance carburetor sizes start at 600 cfm, a 600- or 650-cfm carburetor gives you some wiggle room if you intend to make additional performance modifications later.

Be aware that most manufacturers use a pressure drop of 3.0 inches of mercury (Hg) to rate 2-barrel carburetors and 1.5 in/Hg to rate 4-barrel carbs. These figures assume a maximum achieved vacuum at WOT under full load. In theory, an engine does not achieve a higher vacuum, but in practice they often do and that is generally an indication of the need for the next larger size of carburetor. You can check this in your own vehicle using a manifold vacuum gauge reading at wide-open throttle (WOT) on a

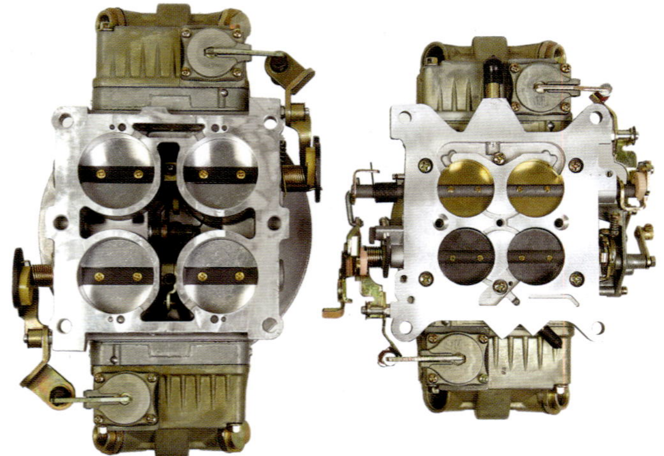

Compare the 2-inch throttle bores on this 1,050-cfm Holley Dominator to the 1⁹⁄₁₆ throttle bores on the adjacent 600-cfm Holley street carburetor. Properly matching carburetor size to actual engine airflow requirements is critical to achieving optimum performance on the street or the track. (Courtesy Holley)

Carburetor venturi size also affects carburetor airflow capacity. Compare the 1¼-inch primary venturis on this 390-cfm Holley carburetor to the 1⅜-inch primaries on a 750-cfm Holley HP 4-barrel. Note the degree of restriction caused by the smaller venturis on the 390 version. (Courtesy Holley)

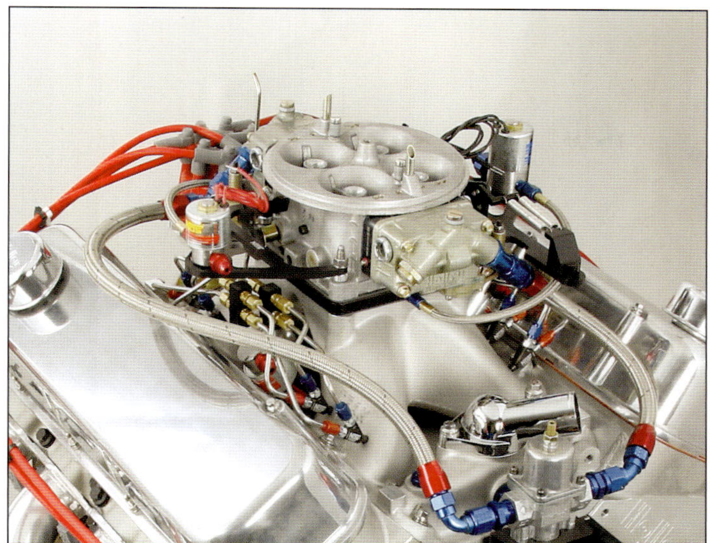

This Dominator is plenty of carburetor for a hot big-block engine, but it would perform poorly on most small-blocks except perhaps larger strokers such as World Products 454-ci small-block Chevy. (Courtesy Holley)

drag strip. If your gauge reads more than 1.5 to 1.7 inches of vacuum under full load in high gear, your carburetor may be too small to deliver maximum performance at peak power.

Race Carburetors

The general rule for racing carburetors is to have 1.1 times the calculated air demand because VE is usually greater than 100 percent. To accommodate the higher efficiency, multiply the theoretical air potential by 1.1. So if our 350 engine is highly modified with racing heads, cam, and intake manifold, we calculate as follows:

$$\text{Race Carb}_{cfm} = [(350 \times 6{,}500) \div 3{,}456] \times 1.1 = 724$$

In this case, you would probably choose a 750-cfm carburetor since it is the nearest common size, especially if you are drag racing. If you are road racing, you might consider a 700-cfm Holley because it has 1/16-inch-smaller venturis that might improve throttle response off of tight corners. In this case the smaller carb may prove to be the better choice. As a general rule, it has been found that going with the smaller carb almost always yields the best results.

Performance carburetors are mostly made in 50-cfm increments, so if your airflow calculations happen to split the difference, choose the smaller carb unless you have a

compelling reason to go larger. Supercharger and turbocharger applications present a higher air demand to handle the increased airflow capacity of the supercharging device. (See "Boost and Supercharger Drive Ratios" on page 72.)

Choosing Throttle Body Size

Throttle body equivalents pretty much parallel carburetor sizing because the engine's airflow requirement is unchanged. In most throttle body applications the air is not burdened with the task of carrying fuel to the valve, but it still needs to maintain sufficient energy (velocity) to support efficient cylinder filling. To calculate an equivalent throttle body size based on known carburetor size, you need to recall the formula for the area of a circle:

$$A = \text{diameter}^2 \times 0.7854$$

Or, in this case:

$$A = \text{diameter}^2 \times 0.7854 \times \text{number of throttle bores}$$

You can use this formula to calculate the throttle bore area of a given carburetor and compare it to an equivalent throttle body. You can calculate the area of a large single throttle body or the combined area of a multiple bore throttle body. Calculate the separate areas and multiply by the number of bores in the carburetor or throttle body.

Here's an example with a 750 Holley carburetor that has four throttle bores measuring $1\frac{7}{16}$ inches compared to a single-bore 75-mm aftermarket throttle body for a fuel injected application:

First, find the total throttle area of the Holley.

Convert to decimals.

$$1\frac{7}{16} = 1.4375 \text{ inch}$$

Find the area.

$$A = 1.4375^2 \times 0.7854 \times 4 = 6.49 \text{ square inches}$$

Calculate the area of the 75-mm throttle body.

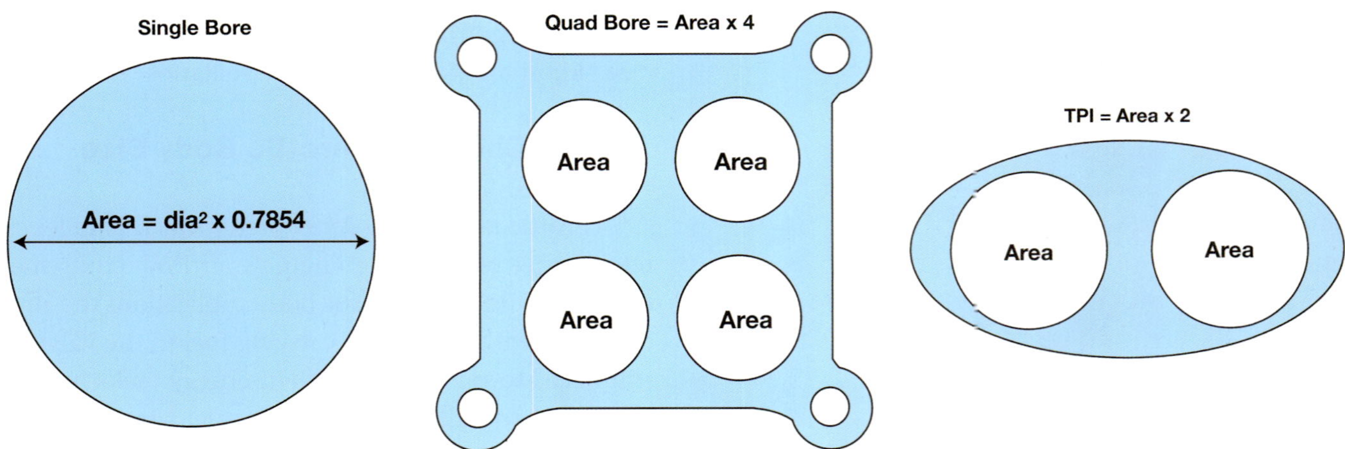

To calculate equivalent throttle bore area, find the area of a single bore as shown here, then multiply it by the number of bores in your carburetor or throttle body.

First convert to inches (multiply 75 mm by the conversion factor 0.0393701).

$$75 \text{ mm} \times 0.0393701 =$$
$$2.9527 \text{ inches (equivalent diameter)}$$

Now calculate the area of the 2.9527-inch-diameter throttle bore.

$$2.9527^2 \times 0.7854 = 6.84 \text{ square inches}$$

Recall that the 750 cfm Holley had 6.94 square inches of throttle area, so the 75-mm throttle body is a little larger. To calculate an exact throttle body equivalent use a shortcut by working in percentages. It will get you very close.

$$6.49 \text{ (Holley)} \div 6.84 \text{ (throttle body)} = 94.8\%$$

$$75 \text{ mm} \times 0.94 = 70.5 \text{ mm}$$

A 70- or 72-mm throttle body would be the closest equivalent for your application.

Calculating Supercharger Carburetor Size

Under boosted conditions, your supercharged engine requires up to 50-percent-more airflow capacity than a

Two appropriately sized Holley 4-barrels ensure an adequate supply of air when the supercharger ramps up its air demand. (Courtesy Holley)

naturally-aspirated engine of equivalent displacement. Not discounting the air cleaner, the carburetor is the primary source of restriction for air entering the supercharger and, ultimately, the engine. The carburetor must be capable of serving the increased airflow demand and the additional fueling requirement under boost. Final air demand depends upon engine characteristics such as manifold and cylinder head efficiency (both VE contributors) and the amount of boost your supercharger is supplying. The calculation is based on engine speed, displacement, and the maximum anticipated boost pressure at WOT.

The following formula calculates the appropriate air demand and corresponding carburetor or throttle body size for your application:

$$CFM_{max} = (displacement \times RPM \div 3{,}456) \times [(max\ boost\ /\ 14.7) + 1]$$

Centrifugal blowers such as this procharger setup come on stronger with increased RPM and compressor speed. Swapping to a smaller pulley on the blower can raise the boost for a little weekend warrior activity. During the week you simply swap back to the larger pulley for daily driving. (Courtesy ATI)

Where:
RPM = max anticipated engine speed
Boost = maximum boost at WOT
14.7 = normal atmospheric pressure at sea level

For example for a 302 Ford making 6 psi boost at 6,000 rpm:

$$CFM = (302 \times 600 \div 3456) \times [(6 \div 14.7) + 1] = 524.3 \times 1.4089 = 738$$

Note that the VE correction factor (0.85) is conspicuously absent from the formula. In theory it shouldn't be because the same restrictions that limit absolute VE in a normally aspirated engine are still present in your supercharged engine, and anything you do to decrease them will aid the efficiency of your supercharger. Most people contend that boost pressure overcomes these restrictions, and in practice it does to a degree, but it still has to perform the work and without the restrictions it will be more efficient.

A similar formula used by many turbo manufacturers also ignores the VE correction. A logical explanation might be that both formulas provide a fudge factor to prevent users from undersizing the carburetor on their engines and

Many nitrous applications also require substantial carburetor support to ensure maximum performance. Even though you are adding extra oxygen chemically, the engine will respond better with larger carburetors. (Courtesy Holley)

thus making the supercharger or turbocharger appear less efficient. While the supercharging device may have enough dynamic range to overcome restrictions, it still introduces undesirable heat to the compressed charge resulting in boost pressure, but not necessarily an increase in air density. If the carburetor acts as a restriction and you drive the blower faster to overcome VE losses, the pressurized air may also gain so much heat that it exceeds the octane rating of your fuel.

VE-corrected formulas calculate the correct air requirement, but you can't go wrong with the uncorrected formula because it ensures that your carburetor will never be a restriction. (If you want to incorporate the VE correction factor into your calculation, simply multiply the result by 0.85 for street carbs.)

Note also that this calculation is for a carburetor used on the inlet side of the supercharger, such as in a Roots-type blower. It is not valid nor should it be used for a blow-through application. Also it should be noted that using this calculation doesn't take into account the fact that Roots blowers are not very efficient and, as such, demand that the inlet side be as unrestricted as possible. That's why you see dual carbs on most Roots blowers unless they are *very* mild street blowers. For a max-effort Roots blower, the inlet side must be left as unrestricted as possible; hence, two very large and oversized carbs. So this math should only be used for single-digit boost applications on mild engines where peak power is not the primary goal.

Calculating Turbocharger Carburetor Size

The recommended method for calculating turbocharger carburetor size is based on a simple air mass calculation related to horsepower. Turbocharger manufacturers use the 10:1 rule which states that it takes one pound of air to make 10 hp. That pound of air varies in density according to the gas laws governing pressure, temperature, and volume, but the 10:1 ratio is relatively constant.

Calculating the air mass in pounds per minute (lbs/min) provides a convenient way of figuring carburetor size. Divide the anticipated HP by 10: 500 hp divided by 10 equals an air mass requirement of 50 lbs/min. The engine must flow 50 pounds of air per minute to make 500 hp.

This 1,600-hp, 299-ci Ken Duttweiler–built Chevy is one angry small-block. Built exclusively for George Poteet's 435-mph Bonneville streamliner, it is a model of mathematical efficiency with a properly sized single turbocharger.

$$\text{Corrected Mass Air Flow} = \text{HP} \div 10$$

$$500 \text{ hp} - 10 = 50 \text{ lbs/min}$$

To convert the calculated air mass to CFM, divide by the standard air density factor 0.0691.

$$\text{CFM} = \text{calculated mass air flow} \div$$
$$\text{standard air density factor (0.0691)}$$

$$\text{CFM} = 50 \text{ lbs/min} \div 0.0691 = 723.6$$

This assumes 100-percent VE so don't forget to correct by the VE correction factor of 0.85.

$$723.6 \times 0.85 = 615 \text{ cfm}$$

If this seems too small it is—because you still have to multiply by the density ratio, which is read from density ratio charts based on efficiency ratios for various turbochargers. In our example we discovered a density ratio of 1.25 for a turbocharger with a 74-percent efficiency ratio.

$$723.6 \times 1.25 = 905 \text{ cfm}$$

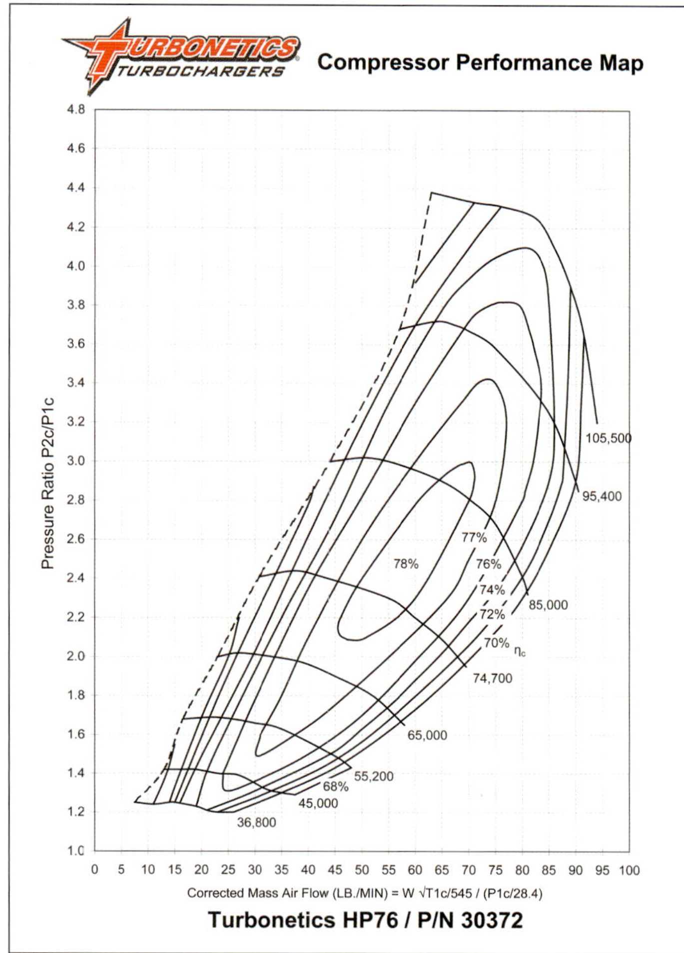

Turbonetics HP76 / P/N 30372

Turbocharger compressor maps are used to select the correct turbocharger based on calculated pressure ratios and the calculated mass air flow in pounds per minute for the horsepower requirement. The efficiency islands on the map indicate various levels of the efficiency range for that particular turbine housing. The point where the pressure ratio and the calculated air mass meet on the map shows the efficiency the turbine will achieve based on those parameters. (Courtesy Turbonetics)

In practice you should consult with the turbocharger manufacturer who will help match your carburetor selection to the corresponding turbocharger based on the turbocharger's pressure ratio and the appropriate air density for the boost you will be running.

Sizing a Turbocharger

The formula for calculating turbocharger boost pressure ratios provides a convenient means of selecting the

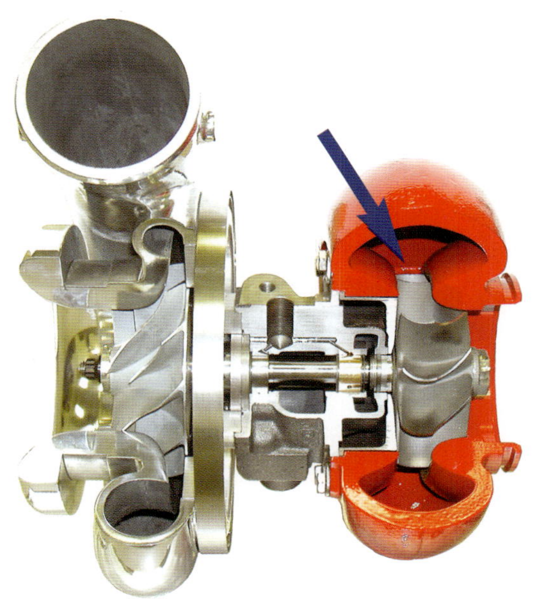

The turbine housing (in red) shows the actual location of the tongue (arrow) where the A/R ratio is measured in the volute. (Courtesy Turbonetics)

The A/R ratio is the square area of the turbine housing volute at the end of the tongue divided by the radius from the center of the turbine wheel to the dynamic center (called the Centroid) of the gas flow path through the volute. (Courtesy Turbonetics)

Understanding the A/R Ratio

The A/R ratio is the area in inches of the turbine housing volute divided by the radius from the center of the turbine wheel axis of rotation to the dynamic flow center of the volute. The volute area is measured at the end of the turbine tongue, which is the point where the exhaust gasses are first exposed to the turbine wheel (not at the foot or flange, but typically several inches down the throat). The radius is taken in the same plane from the turbine wheel center of rotation to the dynamic center of the volute. The dynamic center (or centroid) is the longitudinal axis point where half the flow passes above it and half passes below it. Due to the non-symmetrical slope of the housing, the dynamic center is not the physical center of the volute, but is slightly biased toward the outside of the flow path. This point can only be determined by the turbocharger manufacturer.

The A/R ratio provides a convenient way of ranking turbine size according to a common dimension within the turbine housing. Thus turbine housings can be related to engine displacement and horsepower requirements. Smaller A/R ratios spool up faster, but reach a plateau of efficiency sooner. Larger A/R ratios spool more slowly, but deliver greater efficiency at elevated engine speeds. The dynamics of these relationships are complex and are not something the casual enthusiast can easily calculate. (For a detailed description of A/R ratios and how turbocharger manufacturers apply them, I highly recommend Jay K. Miller's *Turbo: Real World High-Performance Turbocharger Systems*, published by CarTech.)

Your turbocharger manufacturer will help you select the appropriate A/R ratio for your specific application. If your requirements change or you need to alter the response time of the turbocharger, the turbine housing can be swapped to meet the new requirement; switching to a lower A/R for quicker response or to a higher one for greater high speed efficiency. The A/R ratio has a dramatic effect on turbocharger performance, so it's best to let the manufacturer choose it for you based on your stated goals.

The A/R ratio also applies to the turbine or exhaust side of the turbocharger as indicated here by the arrow and oval on the exhaust turbine housing. (Courtesy Turbonetics)

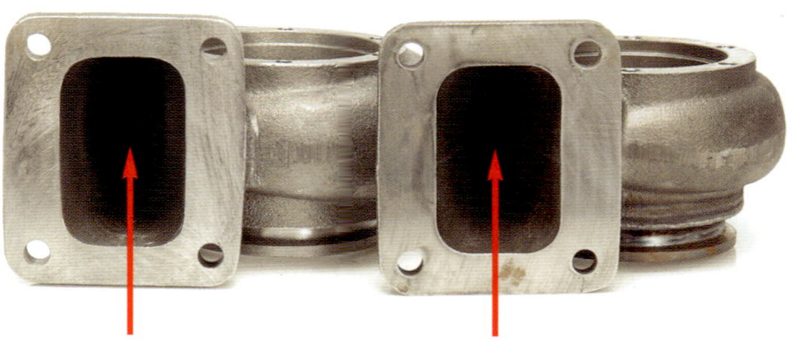

The smaller A/R shown on the right improves turbine response time so the turbo builds boost more quickly. The larger volute on the left is designed to build big power at higher RPM. (Courtesy Turbonetics)

The relationship of the blower pulley (driven) diameter and tooth count to that of the crank (drive) pulley is the blower drive ratio. When the ratio is positive (crank pulley larger), the blower is overdriven. When the crank pulley is smaller the ratio is negative or underdriven and the blower runs slower than crank speed. (Courtesy Weiand)

Modern mini-blowers such as this Weiand system run at higher RPM than traditional Roots blowers, but the drive ratio is still controlled by pulley size. In most cases only the upper pulley is changed to raise or lower the boost. (Courtesy Weiand)

proper turbocharger compressor wheel by plotting the calculated pressure ratio against the previously mentioned mass air calculation in lbs/min on a turbocharger compressor map.

$$\text{Pressure Ratio} =$$
$$\text{max boost} + \text{ambient pressure (usually 14.7)} \div$$
$$\text{ambient pressure (usually 14.7)}$$

Example: If we assume a max boost pressure of 6 psi, we calculate as follows:

$$\text{Pressure Ratio} = (6 + 14.7) \div 14.7 = 1.408$$

Most turbocharger manufacturers publish their compressor maps and density ratio charts online for your convenience. To select an appropriate compressor, you refer to the turbocharger compressor map that plots the corrected mass air flow in lbs/min on the horizontal axis and the calculated pressure ratio on the vertical axis. (The irregular shaped lines in the center that look like a geologist's terrain map are called efficiency islands. They indicate the effective range of a given compressor's efficiency at various pressure ratios and mass air flow levels.)

To use the compressor map you read horizontally from the previously calculated pressure ratio until you meet the vertical line rising from the mass airflow in pounds per minute. The point where they cross will fall somewhere on

Typical Drive Pulleys & Ratios

8mm Pitch Drive Pulleys and Ratios - 2" Registers

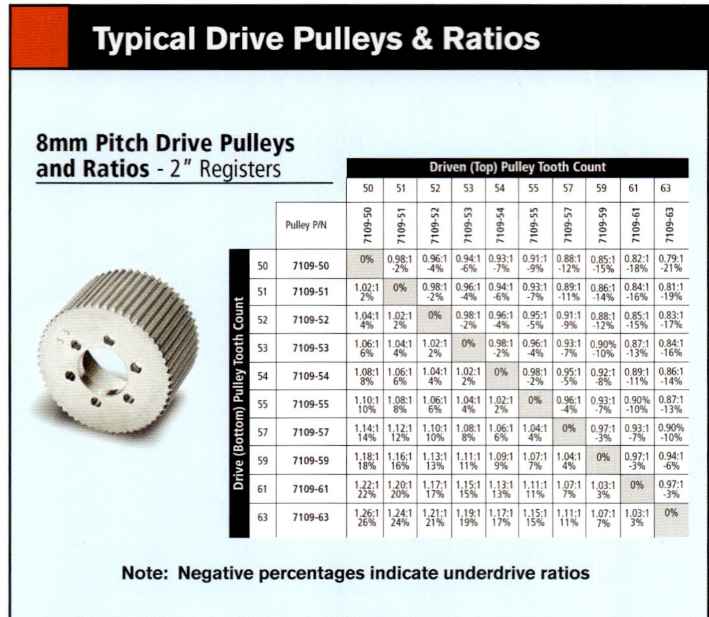

Drive (Bottom) Pulley Tooth Count	Pulley P/N	Driven (Top) Pulley Tooth Count									
		50 7109-50	51 7109-51	52 7109-52	53 7109-53	54 7109-54	55 7109-55	57 7109-57	59 7109-59	61 7109-61	63 7109-63
50	7109-50	0%	0.98:1 -2%	0.96:1 -4%	0.94:1 -6%	0.93:1 -7%	0.91:1 -9%	0.88:1 -12%	0.85:1 -15%	0.82:1 -18%	0.79:1 -21%
51	7109-51	1.02:1 2%	0%	0.98:1 -2%	0.96:1 -4%	0.94:1 -6%	0.93:1 -7%	0.89:1 -11%	0.86:1 -14%	0.84:1 -16%	0.81:1 -19%
52	7109-52	1.04:1 4%	1.02:1 2%	0%	0.98:1 -2%	0.96:1 -4%	0.95:1 -5%	0.91:1 -9%	0.88:1 -12%	0.85:1 -15%	0.83:1 -17%
53	7109-53	1.06:1 6%	1.04:1 4%	1.02:1 2%	0%	0.98:1 -2%	0.96:1 -4%	0.93:1 -7%	0.90% -10%	0.87:1 -13%	0.84:1 -16%
54	7109-54	1.08:1 8%	1.06:1 6%	1.04:1 4%	1.02:1 2%	0%	0.98:1 -2%	0.95:1 -5%	0.92:1 -8%	0.89:1 -11%	0.86:1 -14%
55	7109-55	1.10:1 10%	1.08:1 8%	1.06:1 6%	1.04:1 4%	1.02:1 2%	0%	0.96:1 -4%	0.93:1 -7%	0.90% -10%	0.87:1 -13%
57	7109-57	1.14:1 14%	1.12:1 12%	1.10:1 10%	1.08:1 8%	1.06:1 6%	1.04:1 4%	0%	0.97:1 -3%	0.93:1 -7%	0.90% -10%
59	7109-59	1.18:1 18%	1.16:1 16%	1.13:1 13%	1.11:1 11%	1.09:1 9%	1.07:1 7%	1.04:1 4%	0%	0.97:1 -3%	0.94:1 -6%
61	7109-61	1.22:1 22%	1.20:1 20%	1.17:1 17%	1.15:1 15%	1.13:1 13%	1.11:1 11%	1.07:1 7%	1.03:1 3%	0%	0.97:1 -3%
63	7109-63	1.26:1 26%	1.24:1 24%	1.21:1 21%	1.19:1 19%	1.17:1 17%	1.15:1 15%	1.11:1 11%	1.07:1 7%	1.03:1 3%	0%

Note: Negative percentages indicate underdrive ratios

The chart above is for a Weiand 6-71 blower with 8-mm-pitch drive pulleys. It illustrates how one pulley can be changed to achieve a specific blower speed.

Effective Compression Ratio Chart

Static Compression Ratio	Pump Gas Blower Boost Pressure (psi)							Race Gas Blower Boost Pressure (psi)					
	2	4	6	8	10	12	14	16	18	20	22	24	26
6.0:1	6.8:1	7.6:1	8.4:1	9.3:1	10.1:1	10.9:1	11.7:1	12.5:1	13.3:1	14.2:1	15.0:1	15.8:1	16.6:1
6.5:1	7.4:1	8.3:1	9.2:1	10.0:1	10.9:1	11.8:1	12.7:1	13.6:1	14.5:1	15.3:1	16.2:1	17.1:1	18.0:1
7.0:1	8.0:1	8.9:1	9.9:1	10.8:1	11.8:1	12.7:1	13.7:1	14.6:1	15.6:1	16.5:1	17.5:1	18.4:1	19.4:1
7.5:1	8.5:1	9.5:1	10.6:1	11.6:1	12.6:1	13.6:1	14.6:1	15.7:1	16.7:1	17.7:1	18.7:1	19.7:1	20.8:1
8.0:1	9.1:1	10.2:1	11.3:1	12.4:1	13.4:1	14.5:1	15.6:1	16.7:1	17.8:1	18.9:1	20.0:1	21.1:1	22.1:1
8.5:1	9.7:1	10.8:1	12.0:1	13.1:1	14.3:1	15.4:1	16.6:1	17.8:1	18.9:1	20.1:1	21.2:1	22.4:1	23.5:1
9.0:1	10.2:1	11.4:1	12.7:1	13.9:1	15.1:1	16.3:1	17.6:1	18.8:1	20.0:1	21.2:1	22.5:1	23.7:1	24.9:1
9.5:1	10.8:1	12.1:1	13.4:1	14.7:1	16.0:1	17.3:1	18.5:1	19.8:1	21.1:1	22.4:1	23.7:1	25.0:1	26.3:1
10.0:1	11.4:1	12.7:1	14.1:1	15.4:1	16.8:1	18.2:1	19.5:1	20.9:1	22.2:1	23.6:1	25.0:1	26.3:1	27.7:1
10.5:1	11.9:1	13.4:1	14.8:1	16.2:1	17.6:1	19.1:1	20.5:1	21.9:1	23.4:1	24.8:1	26.2:1	27.6:1	29.1:1
11.0:1	12.5:1	14.0:1	15.5:1	17.0:1	18.5:1	20.0:1	21.5:1	23.0:1	24.5:1	26.0:1	27.5:1	29.0:1	30.5:1

This chart shows the static and effective compression ratio at various boost levels for both pump gas and race gas. The line through the center delineates the practical limit for most pump-gas applications. (Courtesy Weiand)

you've done your calculations carefully, you might even be surprised to find that they approve of your preliminary selection.

one of the indicated efficiency islands. For a street application you want a minimum of 65- to 70-percent efficiency and it's not usually difficult to find a compressor that will put you in the 72- to 74-percent range.

Here again it is fun to brainstorm compressor maps and efficiency ratios, but for best results, work with your turbocharger manufacturer to select the best turbine for your particular application and anticipated usage. If

Boost and Supercharger Drive Ratios

The amount of boost your supercharger delivers depends on its size and efficiency and the speed that you drive it relative to the crankshaft. The crankshaft is the "drive" or driven device and the supercharger is the "driven" device. Your ability to adjust supercharger boost is controlled by the supercharger drive ratio, which is the relationship between the size and tooth count on the drive pulley relative to that of the blower or driven pulley. Before

Weiand 6-71 Drive Ratio & Estimated Boost Chart (psi)

	1.30:1	1.25:1	1.20:1	1.15:1	1.10:1	1.05:1	1:1	0.95:1	0.90	0.85:1	0.80:1	0.75:1	0.70:1
Engine	30%	25%	20%	15%	10%	5%	0%	-5%	-10%	-15%	-20%	-25%	-30%
327	27.1	25.5	23.9	22.3	20.7	19.1	17.5	15.8	14.2	12.6	11.0	9.4	7.8
350	24.3	22.8	21.3	19.8	18.3	16.8	15.3	13.8	12.3	10.8	9.3	7.8	6.3
383	21.0	19.6	18.2	16.9	15.5	14.1	12.8	11.4	10.0	8.6	7.3	5.9	4.5
392	20.2	18.8	17.5	16.1	14.8	13.5	12.1	10.8	9.4	8.1	6.8	5.4	4.1
400	19.5	18.2	16.8	15.5	14.2	12.9	11.6	10.3	9.0	7.6	6.3	5.0	3.7
454	15.4	14.2	13.1	11.9	10.8	9.6	8.5	7.3	6.1	5.0	3.8		
502	12.5	11.5	10.4	9.4	8.3	7.3	6.2	5.2	4.1	3.1			
540	10.6	9.6	8.7	7.7	6.7	5.7	4.8	3.8					

This chart shows estimated boost levels for common drive ratios and engine sizes. (Courtesy Weiand)

you consider increasing the boost, keep in mind that higher boost levels do not necessarily produce denser air for the engine to process. Depending on your combination, it may simply result in hotter air at higher pressure but no real increase in useful oxygen content. For this reason large increases in boost are ill advised unless you can supplement them with effective charge cooling or higher octane fuel.

Supercharger speed is determined by the blower drive ratio. If the drive pulley and the driven pulley both have the same tooth count, the drive ratio is 1:1 and the blower turns at the same speed as the crankshaft. If the crank (drive) pulley is larger than the blower (driven) pulley, the blower is overdriven by the relative percentage. If the blower pulley is larger than the crank pulley the blower is underdriven. Use the following formulas to calculate the percentage of over- or underdrive based on pulley size (tooth count):

$$\% \text{ Overdrive} =$$
$$\text{tooth count, driven pulley (blower)} \div$$
$$\text{tooth count, drive pulley (crank)}$$

$$\% \text{ Underdrive} =$$
$$\text{tooth count, drive pulley (crank)} \div$$
$$\text{tooth count, driven pulley (blower)}$$

To estimate supercharger speed at any given engine speed, multiply the engine speed by the current blower drive ratio.

$$\text{Supercharger Speed} = \text{RPM} \times \text{blower drive ratio}$$

If you have a supercharger pulley arrangement where the blower is overdriven by a ratio of 1.1 (10 percent) and the maximum engine speed is 6,500 rpm, the blower speed is 7,150 rpm.

$$\text{Supercharger Speed} = 6,500 \times 1.1 = 7,150 \text{ rpm}$$

Supercharger manufacturers publish their supercharger drive ratios and corresponding boost levels in charts that show the tooth count and percentage of over- or underdrive. These charts indicate the amount of boost you can expect from a given ratio relative to the displacement of your engine.

For example, a Weiand 6-71 blower driven 1:1 delivers 8.5 pounds of boost on a 454-ci engine. You can raise that to 9.6 pounds by overdriving the blower 5 percent (1.056 times crank speed). The accompanying charts indicate the appropriate pulley relationship to achieve the desired boost level.

If you need to lower the boost to avoid detonation, you can underdrive the blower 5 percent (0.95 times crank speed) to get 7.3 pounds. Usually you only have to change one pulley, or swap upper and lower pulleys to achieve your goal.

Overdrive percentages on the older style 6-71 and 8-71 blowers rarely exceed 30 percent, but note that the more recent mini-blowers and centrifugal blowers are driven much faster; more than double crank speed in some cases. You have to pay careful attention to these blowers, especially the centrifugal type, because they can easily exceed boost limits at higher engine speeds. Consult the manufacturer's published drive ratios and boost charts to match your particular application.

Wave Tuning

We tend to think of intake flow as the smooth, steady passage of an air/fuel mixture through an intake port and into a cylinder, but the dynamics are far more complicated. Engineers have long recognized the effects of wave tuning within the inlet system, but until recently it was never deemed necessary or cost effective to pursue it in production vehicles. Racers have known about it for a long time as evidenced by the racing efforts of the Ramchargers in the late 1950s and early 1960s. Various racing efforts have made good use of wave tuning over the years.

What we find in the inlet tract is a continuous series of starts and stops as the intake valve opens and closes. Within this physical stream of flow, supersonic pulses are reflected back and forth between the cylinder and the atmospheric source of inlet flow. These are high- and low-pressure pulses depending on the point of origin. The intake flow has inertia, which creates a low-pressure pulse reflected from the intake valve every time it opens. This reverse pulse travels back toward the inlet entry much faster than, and right through, the still onrushing gas mixture until it reaches the inlet entry or atmospheric pressure. At that point a positive high-pressure pulse is reflected back toward the valve. The reflected pulse travels down the inlet tract until it reflects off the piston top as another high-pressure pulse traveling back toward the entry again. This time the inlet entry reflects it back as a low-pressure pulse and the cycle begins again. Within this series of alternating high-speed pulses we can make good use of the incoming high-pressure pulses to increase the density of the fuel charge by using its energy to effectively boost more molecules into the cylinder.

Dyno tests with individual runner (IR) manifolds have shown significant power gains when the lengths of the intake stacks are carefully matched to take advantage of pulse timing. The trick is to match camshaft timing and inlet tract length so that a high-pressure pulse arrives just as the valve opens and sweeps additional mixture into the cylinder.

The speed of a pulse through an air/fuel mixture varies with its temperature, but for calculation purposes it is most often related as 1,100 ft/sec at 100 degrees F, which is representative of real-world temperatures. Pulse timing is controlled by the speed of the pulse, the length of the inlet path, and the timing of intake valve event. Since valve operation and pulse speed are fixed, the length of the inlet can be tuned to achieve resonance at some particular engine speed that we can take advantage of.

In the 1960s, Chrysler engineers established a mathematical constant (K-value) which enabled them to calculate the optimum intake length within a range of plus-or-minus 3 inches. While that seems a bit vague, it is also known that some benefit begins to accrue on either side as engine speed approaches the "sweet spot." This point of resonance is calculated as follows:

$$L = [(K \times C) \div N] \pm 3$$

Where:
L = length of the inlet path in inches
K = mathematical constant (Chrysler chose 72)
C = speed of the pulse (arbitrary according to temperature)
N = engine speed (RPM)
± = recommendation to encourage experimentation to determine the actual "sweet spot"

So for a given engine speed of, say, 6,000 rpm, we can calculate an optimum inlet path length.

$$L = [(72 \times 1,100) \div 6,000] \pm 3$$

$$L = 12 \pm 3 \text{ inches}$$

This would be the ideal length from the intake valve to the inlet entry for atmospheric pressure. Performance author Phillip Smith addresses this concept in his book *Scientific Design of Exhaust and Intake Systems*. His research derived a K-value of 90, which substantially increases the length requirement.

In the absence of more precise research (which undoubtedly exists within the engineering departments of major automakers), it is difficult to determine the ideal K-value. We know from experience that critically-tuned inlets deliver exceptional power when operated within the narrow range of their optimum engine speed. If a car's transmission gears are selected to provide minimal RPM drop on each shift, the engine can be run within its peak efficiency range most of the time. This is reinforced by factory efforts to investigate and implement variable-length inlet systems, and by the fact that almost all OEM performance engines make use of very precise inlet-length tuning like that found on third-generation GM LS series small-blocks and recent Chrysler Hemis.

It is more difficult to apply this in practice because aftermarket intakes are manufactured to a fixed length, but you might think about why tunnel ram manifolds are so effective at high-RPM ram tuning, particularly above 7,000 rpm. Take a look at any Pro Stock intake (if it's not covered up) and you'll see runners optimized for the high-RPM Pro Stock engine environment. The higher the RPM, the shorter the runner. Another critical factor is runner taper (typically about 4 degrees), which is easily seen on most manifolds. The use of taper is beyond the scope of this book, but you can learn all about it with Motion Software's Dynomation program (download the manual for free at www.motionsoftware.com), which makes extensive use of wave tuning and taper for accurate engine simulation.

To some degree this thinking is partially evident in street intakes, like the high-torque Edelbrock Performer RPM, which employs lengthy constant cross-sectional area runners to boost torque. The runner length is, by necessity, an ideal compromise, but manufacturers have tried very hard to size these manifolds for optimum results knowing full well that they will be employed across a broad range of engine sizes and speeds. Their performance proves that they have done an exceptional job. On the other hand, if you're running a 5-mile-long, high-RPM pull at Bonneville with an IR intake system, critical dyno testing can help you pinpoint the exact length to cut your stacks for optimum wave tuning in your desired RPM range. It's food for thought and kind of fun, too.

Ram Effects and Inlet Cooling

Many racers make use of a hood scoop to capture and direct more air into the carburetor(s). If properly configured with a sealed air box, it can provide a slight pressure boost that can increase power. It may also provide additional power due to the cooling effect of the high-speed air entering the air box.

First, consider the "ram air" effect of high-speed air entering the scoop and the air box. A good system has a sealed air box surrounding the carburetor entry. Depending on the layout, the scoop is often designed with a smaller opening that expands into the air box to help slow the air so it can make the turn into the carburetor(s). It's well known that the shape of the scoop entry determines the degree of efficiency it provides. Scoops with straight cut edges are not very effective, but those with a generously curved radius can recover as much as 90 percent of the incoming air pressure, which of course increases with the car's velocity. The following formula is used to calculate the pressure increase provided by ram air assuming a nicely curved entry radius and 90 percent pressure recovery.

$$P_{vel.} = (V^2 \times p) \div 4{,}311$$

Where:
$P_{vel.}$ = velocity-induced pressure increase
p = standard air density
V = vehicle speed in mph
4,311 = mathematical constant

If we assume standard air density of 0.0691 we can calculate the pressure increase according to the speed of the vehicle. In this case we'll assume a drag car that achieves 165 mph by about the 1,000-foot mark.

$$P_{vel.} = (165^2 \times 0.0691) \div 4{,}311 = 0.436 \text{ psi}$$

That assumes perfect pressure recovery, which is virtually impossible, so we'll add a pressure recovery factor of 90 percent, assuming our scoop has a well-rounded inlet radius.

$$P_{vel.} = (165 \times 0.0691 \times 0.90) \div 4311 = 0.392 \text{ psi}$$

That's about four tenths of a pound positive pressure at the carburetor. To calculate the percentage of power increase it might provide, we can refer to standard atmospheric pressure:

$$\% \text{ increase} = (14.7 + 0.392) \div 14.7 = 1.026$$

That's roughly 2½ percent, or 15 hp, on a 600-hp car. To quantify this you could use a data acquisition system with a pressure sensor located in the scoop. Since the system also records time and speed, you can log the scoop's pressure recovery efficiency throughout an entire run as speed and air pressure increase.

Depending on engine speed, air temperature, and the efficiency of your induction system, you may also achieve a slight cooling effect by increasing velocity through the carburetor. If, through data logging you can determine that the incoming air temperature is 100 degrees F and the temperature in the manifold plenum drops to 88 degrees F due to velocity cooling, you can calculate the potential power gain provided by the denser air. To do so, you have to add the recorded temperature to degrees Kelvin.

$$V_1 = (\text{plenum temp} + 459.4) \div (\text{inlet temp} + 459.4)$$

$$V_1 = (88 + 459.4) \div (100 + 459.4) = 0.978$$

To convert the new air density, divide 1 by 0.978.

$$1 \div 0.978 = 1.02$$

So, the cooling effect could supply 2-percent-denser air. And since power is related to air mass flow per minute, you can expect a power increase of roughly 2 percent from inlet cooling. That's an additional 10 to 12 hp on a 600-hp engine.

Again, it is difficult to quantify all of this without accurate data logging to compare against track performance, which would most easily be seen as increased trap speed. To further evaluate it you may want to plug off the scoop or run a flat hood if possible so the engine breathes underhood air temperature with no pressure increase. And you also have to consider the aerodynamic drag of the scoop, which may offset some of the gain from ram air pressure and airspeed cooling. It's always something!

How to Calculate Runner Cross Section

As noted earlier, intake runner cross section is a primary influence of torque peak location within the engine's power band. If you're inclined to calculate such things it can be difficult because manufacturers do not publish this information; indeed, some of them may not even know it themselves. It's well known that Edelbrock's Performer RPM dual-plane intake features constant cross-section runners. And there is no doubt that Edelbrock has sized the runners for optimum performance across the normal street operating range. The constant cross section ensures consistent runner velocity, which discourages air/fuel separation. Because air and fuel molecules differ in weight, they are susceptible to separation when the mixture experiences velocity changes according to variations in cross section or abrupt changes in direction. Edelbrock's diligence in this regard is a primary reason why the Performer RPM is one of the most potent and popular dual-plane intakes available.

Still, those who wish to determine the average cross section of an unknown intake runner can achieve a close approximation by carefully measuring the dimensions of the runner inlet in the plenum and the runner exit at the manifold/cylinder head flange. Using the appropriate size telescoping gauge (snap gauge), measure the runner inlets at a point far enough down the runner to avoid a false reading at the rounded runner entry. Calculate the cross section there and at the runner exit and average the two to get the mean cross section of the runner. This is sufficient for calculating a torque peak value for the purpose of comparison with other intakes and is, of course, independent of any air/fuel separation issue that may exist in some manifolds due to sharp runner turns or inconsistent velocity caused by area variations.

Knowing the cross section allows you to predict the torque peak RPM, but you still can't adjust it with a fixed manifold casting. You can partially influence it to the degree that you can broaden it via complementary adjustments to the header primary pipe cross section, which is easier to change. Calculate the torque peak RPM:

$$\text{Torque Peak RPM} = \text{runner cross section} \times 88{,}200 \div \text{individual cylinder volume}$$

Then reverse the formula to calculate a header cross section that promotes a secondary peak to broaden the curve.

$$\text{Cross Section} = \text{Peak torque RPM} \times \text{cylinder volume} \div 88{,}200$$

While not exact, this method can help you ballpark a header selection that effectively complements your fixed area intake manifold. If you wish to pursue it further, consider the difference in intake runner lengths within the same manifold. Since runner length tends to rock the torque curve about the torque peak, you may be able to use changes in selected primary pipe header lengths to complement individual runner lengths in the manifold.

To calculate the average intake-runner cross section, measure the height and width of the runner entry and exit with dial calipers. The entry and exit are usually different so take the average of the two cross sections.

Induction Formulas at a Glance

Potential Air Capacity$_{cfm}$ = displacement x RPM ÷ 3,456

VE % = (measured CFM ÷ potential CFM) x 100

Street Carb Size$_{cfm}$ = displacement x RPM ÷ 3,456 x 0.85

Race Carb Size$_{cfm}$ = displacement x RPM ÷ 3,456 x 1.1

Alternate CFM Formula = (Ci ÷ 2) x RPM x 0.85 (street) or 1.1 (race) 1,728

Supercharger Carb Size$_{cfm}$ = (displacement x RPM ÷ 3,456) x (maximum boost ÷ 14.7 +1)

Turbocharger Air Requirement$_{lbs/min}$ = estimated HP ÷ 10

    or Corrected Mass Air Flow$_{lbs/min}$ = HP ÷ 10

Turbocharger Carb Size$_{cfm}$ = calculated mass air $_{lbs/min}$ ÷ standard air density factor

    (Multiply by 0.85 to incorporate VE correction.)

Turbocharger Pressure Ratio = (maximum boost + ambient pressure) ÷ ambient pressure

Percentage Blower Overdrive = tooth count, upper pulley ÷ tooth count, lower pulley

Percentage Blower Underdrive = tooth count, lower pulley ÷ tooth count, upper pulley

Area of Metric Throttle Body = diameter in mm x 0.0393701 x 0.7854 x number of throttle bores

Throttle Body Diameter If Area Is Known = $\sqrt{\text{calculated throttle body area} ÷ 0.7854}$

Supercharger Speed = RPM x blower drive ratio

Tuned Inlet Length = [(72 x 1,100) ÷ RPM] ± 3 inches

Velocity Pressure Increase = (vehicle speed^2 x 0.0691 x 0.90) ÷ 4,311

Percent Power Increase = (14.7 + velocity pressure increase) ÷ 14.7

Velocity Cooling % Gain = (plenum temperature + 459.4) ÷ (inlet temperature + 459.4)

% Air Density or Power Gain = 1 ÷ velocity cooling % gain

Torque Peak RPM = runner cross section x 88,200 ÷ individual cylinder volume

Cross Section = peak torque RPM x cylinder volume ÷ 88,200

CYLINDER HEAD MATH

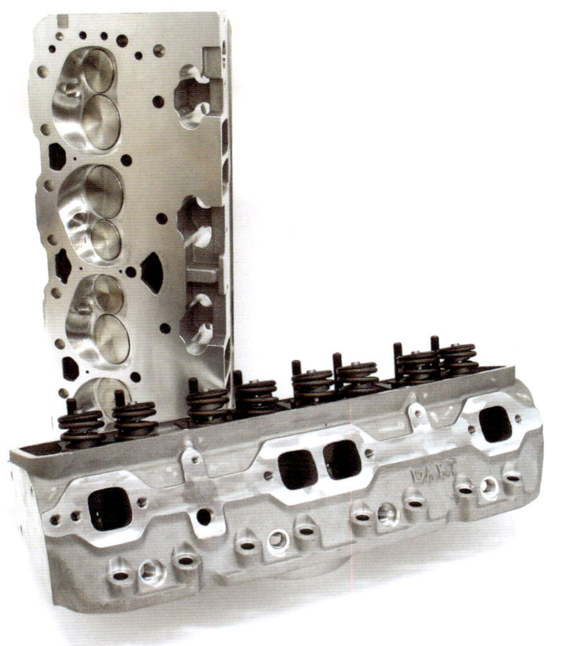

Next to power adders, performance cylinder heads are the most powerful investment you can make. Take careful note of the formulas in this chapter. They will help you understand and evaluate the broad selection of cylinder heads available for your performance project.

Cylinder heads involve a lot more than just rocker arm ratios and valve spring installed heights. In this chapter I discuss formulas for converting combustion chamber volumes, exhaust to intake ratios, valve curtain area, port cross-sectional area, and various other factors that affect the performance potential of any cylinder head. You may see some new stuff here, but none of it is difficult, and you may actually find it fun to brainstorm various combinations that might apply to your particular performance project.

Converting Combustion Chamber Sizes

Cylinder head combustion chambers are cc'd for the purpose of calculating compression ratios and to verify equal volume in each chamber. For the compression ratio formula (see Chapter 3) you need to convert measured cubic centimeters to cubic inches. There are several conversions to choose from.

$$\text{Chamber size} = \text{measured cc} \times 0.0610237$$

That's a lot of numbers to remember and a lot of keys to punch on your calculator, so most engine builders use the following alternate formula.

$$\text{Chamber size} = \text{measured cc} \div 16.4$$

The exact conversion is 16.387064, but the difference is negligible and won't normally affect your final calculation. Check out the following examples using all three versions of the conversion factor to calculate the size (in cubic inches) of a 64-cc combustion chamber.

$$64 \times 0.0610237 = 3.9055 \text{ ci}$$

$$64 \div 16.4 = 3.9024 \text{ ci}$$

$$64 \div 16.387064 = 3.9055 \text{ ci}$$

Note that 16.4 is a rounded number that is easiest to keep in your head. In practice the difference is so slight that it won't affect your compression ratio calculation, so most people opt for the shortcut.

Evaluating Port Volumes

There are no real formulas to work here, but quantifying the difference among cylinder head port cubic centimeters is important if you have performed any porting or cleanup work in the ports. Most performance cylinder heads have published volumes that are usually pretty accurate. You can verify them by cc'ing each port the same way you do a combustion chamber. Install an intake valve using a light checking spring and retainer to hold the valve closed. Many heads have a hole in the port roof that has been drilled and tapped to accept a rocker stud above the port. For accuracy, you should install the rocker stud and its pushrod guide plate (if there is one) to plug the hole to the correct depth. Then cc the port as described in Chapter 3. Since intake ports hold considerably more volume than combustion chambers, it is helpful to have a graduated burette with greater capacity, say, 250 cc, if possible. Otherwise you have to stop the flow of checking fluid at zero and refill the burette one or more times to complete the job. To promote equal work from each cylinder, you want to ensure equal port volumes, and that's why you have to check them if you have done any work in the port or valve bowl area.

Many street engine builders like to clean up the roughness in the bowl area just above the valve and match the port openings to the intake manifold, but they take great care not to alter the cross-sectional area of the valve throat venturi where the rectangular or oval-shaped port makes the transition to a circular shape just above the valve seat. Minor blending into the valve seat is all

This cutaway port shows the extent of the volume you are measuring. Check the port roof for open rocker stud holes that may exist in your particular casting. Plug them with a stud and sealer prior to cc'ing. Be sure to include the guide plate to position the stud depth correctly.

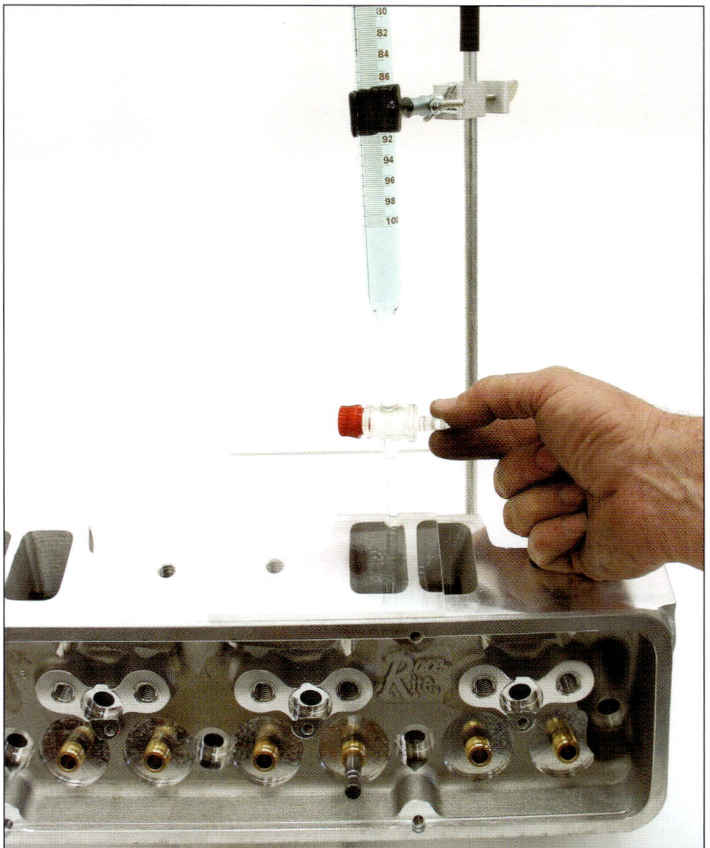

Measure port volume the same way you measure chamber volume (see Chapter 3). Since port volumes are often three or more times chamber volume, you may want to consider a graduated burette with more capacity. Chemical supply houses on the Internet are a good source.

How to CC Combustion Chambers

For maximum performance, all combustion chambers should be of equal volume. This promotes an equal amount of work in the combustion chamber by standardizing the compression ratio.

Individual chamber volumes should be checked by filling each chamber with a measured amount of checking fluid such as alcohol tinted with food coloring. The cylinder heads don't have to be assembled for this procedure, but the valves must be installed with a light coat of grease on the seats to ensure a leak-free seal. You also need to install the correct spark plug. Don't substitute a checking plug unless it is identical to the plugs you will run in the engine, i.e., standard tip or projected tip with the correct plug reach dimension. Ideally, the heads should be supported with cylinder head stands, but lacking those, you can substitute props such as a 2 x 4 block or some other support that facilitates tipping the head slightly so that the deck surface is not quite level. This makes it easier to fill the chambers and lessens the chance of air bubbles that can affect the accuracy of you measurement. You can even bolt a cylinder head to your engine stand and rotate it to just the right position.

To seal the chamber surface, you'll need a 6 x 6 x 1/4-inch clear Plexiglas plate with 1/4 inch hole offset several inches from the center point toward one edge. A thin coating of light grease is carefully smeared around the edges of the combustion chamber and the plate is pressed down over the chamber with the filling hole located close to the edge of the chamber on the high side. Take care that none of the grease is squeezed into the chamber; it could affect the accuracy of you measurements. To achieve accurate measurements you need a graduated burette such as those found in commercially available cylinder head measuring kits sold by www.summitracing.com. These burettes are graduated in tenths of a cubic centimeter (cc).

Once you have the head prepared, carefully fill the burette to the top line. Capillary action will cause the fluid to climb the inner walls of the glass tubing slightly, so make sure you measure to the same point (usually the top or bottom) of the indicated level. Position the burette over the cylinder head, aligning the tip so that the checking fluid can be bled into the chamber through the filling hole in the plate without touching the sides of the opening. Open the petcock slowly and allow the fluid to flow into the chamber. Don't let it fill too quickly as the fluid may splatter out and alter the accuracy of your measurement. Since the head is tilted, the fluid will fill until it touches the plate and then begins working its

Install both valves in the combustion chamber using a light coating of white grease on both of the valve seats to ensure a leak-proof seal. Do not use checking valves as they won't have exactly the same shape and volume displacement as the specific valves you intend to use.

Install the correct spark plug for your application. If you're going to run a projected tip plug or one of the newer-style plugs with a multi-strap ground electrode, use the same type of plug to simulate the volume displacement of the exact plugs you intend to run, including indexing washers (if used).

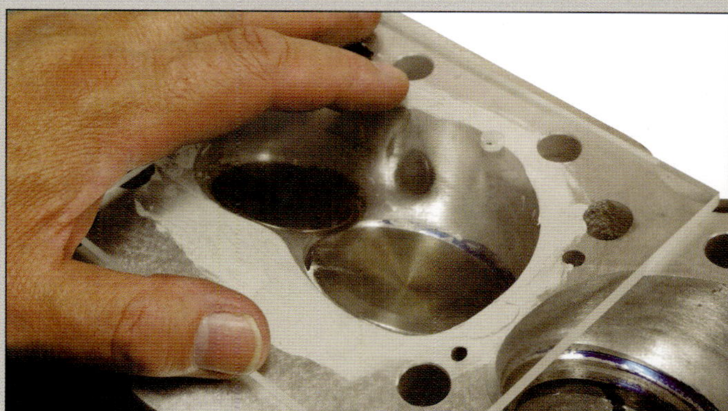

Tilt the head so one side of the chamber is higher than the other. Install the clear plate with the hole located close to chamber edge toward the high side. Use light grease around the perimeter to seal the plate.

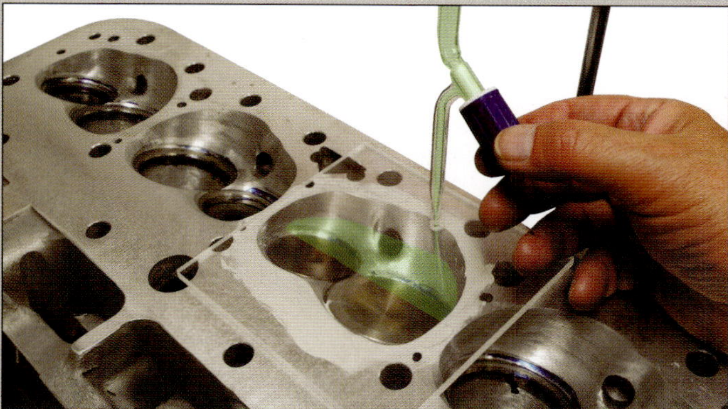

Fill the checking burette to the top of the graduated fill marks. Then position the burette over the chamber, aligning the feed tip with the filling hole. Open the petcock and slowly fill the chamber with checking fluid.

way up toward the filling hole. As it approached the filling hole, turn the petcock to slow the rate of filling. Let it slowly fill until the fluid just reaches the bottom of the filling hole indicating that the chamber is full. Close the petcock valve quickly to avoid fluid coming out of the filling, hole. While filling watch the edges of the chamber and the appropriate intake and exhaust ports in case fluid attempts to leak past the plate seal or the valves.

Record the volume of each chamber. Then determine the best method for equalizing them. A common practice is to mill the head with the largest chamber to match the largest chamber on the other head. The remaining chambers are then worked over until they are equal in volume. Just remember that you should never sink a valve deeper in its seat to gain chamber volume. That destroys the flow capacity of the valve.

If your heads are being used for class competition where no grinding is allowed, you're pretty much stuck with a basic mill cut, but heads for other applications, including street engines, can be dialed-in with minor grinding in the chambers. Note that this requires keeping your burette equipment set up until you verify the results of you milling and grinding. You can use this same method to cc intake ports. Just take care to seal the valve with light grease and install any rocker arm studs that may plug a hole in the roof of the port.

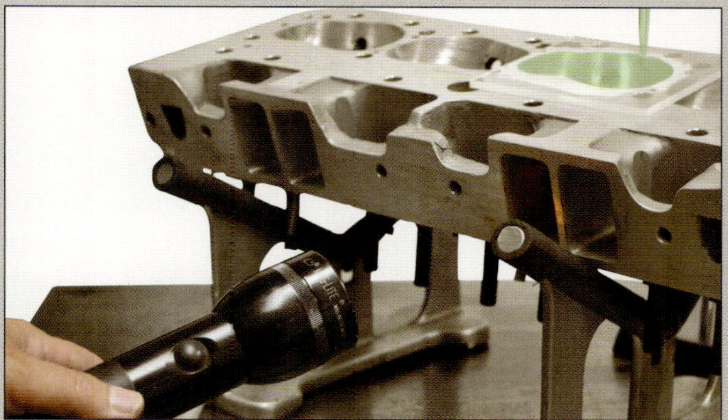

Check for leaks. After the chamber has been filled observe it closely for a few seconds to make certain the level does not fall. If it does, you have a leak at the valves, the spark plug, or the filling plate. Even with the slightest leak, you should redo the measurement to verify accuracy.

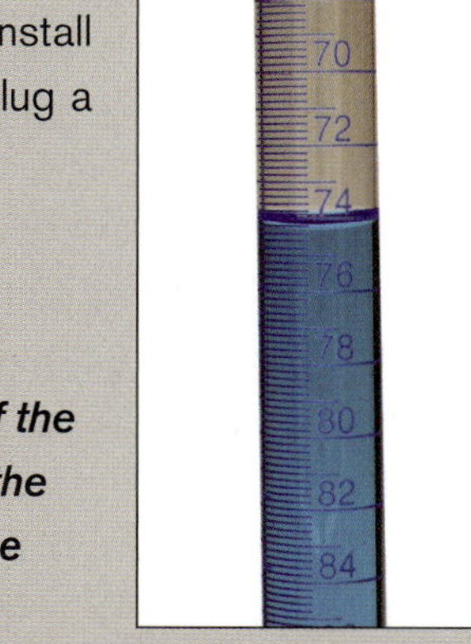

Read either the top or bottom of the fluid level. First set the level on the burette to zero and then read the chamber volume when it is full.

that's required. Altering the area without knowledge and experience can ruin a good port and you would never know it without comparison flow bench work. A better choice for most DIY builders would be to verify what you've already got by cc'ing all the ports and comparing them by percentage.

% port difference = cc larger port ÷ cc smaller port x 100

Compare the following measured ports on a typical performance cylinder head that has been bowl ported and intake matched.

	Cylinder 1	Cylinder 2	Cylinder 3	Cylinder 4
cc	190.2	190.8	192.3	191.0
%	100	100.3	101.1	100.4

All four ports are close and the minor differences are probably not enough to matter for most street applications. The larger difference on cylinder 3 is likely caused by an attempt to match the intake runner that required the removal of more material than anticipated. In most cases you would let it slide, but it can become an issue on certain applications such as a big-block Chevy where you are dealing with the old good port/bad port scenario. In this case you want to be extra careful not to harm the velocity characteristics of the bad ports. It is problematic because anything you do will likely increase volume and impact port velocity. The results can only be verified on a flow bench, which adds to your expense. For most street applications you can probably accept the as-cast port volumes with maybe a slight cleanup as long as port volume difference is held to about 1 percent.

Calculating Valve Curtain Area

When contemplating engine combinations and camshafts in particular, it is often useful to calculate the valve curtain area for a given valve lift and compare it to proposed changes by percentage. The valve curtain area is the area of the flow window that opens when the valve is lifted off its seat. Say you have a 2.02-inch intake valve that opens to 0.500-inch lift. What is the valve curtain area and how much will it increase if you open the valve to 0.535 inch?

To perform the calculation correctly you can't go by the valve diameter itself. You have to go by the flow diameter, which is where the actual valve seat begins. This is usually about 0.040-inch smaller than the average valve diameter. For most calculations it's pretty accurate to simply multiply the valve diameter by 0.98. The calculation then becomes the valve diameter times 0.98 times pi times the lift value. The result is the total available flow area for the valve's flow diameter at the given valve lift.

Valve curtain area =
valve diameter x 0.98 x 3.14 x valve lift

or per our example:

2.02 x 0.98 x 3.14 x 0.500 = 3.107 square inches

To find the percentage of change divide the new valve lift by the current valve lift or do the same with the calculated valve curtain areas.

% = 0.535 ÷ 0.500 = 1.07,
or a 7-percent increase in valve lift

Now the valve curtain areas:

2.02 x 0.98 x 3.14 x 0.500 = 3.107 square inches
2.02 x 0.98 x 3.14 x 0.535 = 3.325 square inches

% = 3.32 ÷ 3.10 = 1.07,
or a 7-percent increase in available flow area

It is important to recognize that this does not represent a 7-percent increase in flow (you wish), but rather a 7-percent increase in potential flow area. Flow gains are still dictated by the combination of available flow area, port velocity and cross-sectional area, valve job, opening rate, and other factors influencing the induction system. The additional flow area relative to valve opening rate and duration offers increased potential for overall cylinder filling.

Increasing the valve lift to 0.550 yields a 10-percent increase in valve lift and available flow area. Again, it does not mean a 10-percent increase in airflow, but rather a 10-percent increase in flow area and, thus, flow potential. Any actual increase would have to be verified on a flow bench.

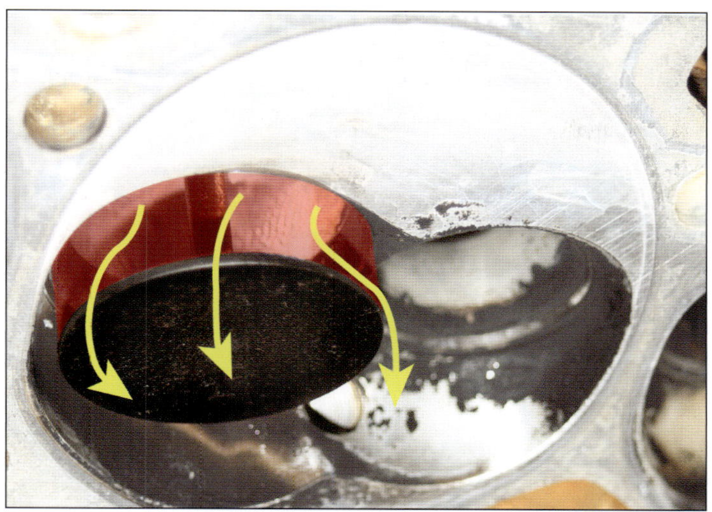

The open valve creates a flow window or so-called valve curtain that provides flow area according to its circumference at the flow diameter times the amount of total valve lift. Calculate it using the diameter of the outer edge of the valve seat, not the overall diameter of the valve.

This cutaway view shows the valve curtain area so you can visualize the flow window of the open valve.

Another important thing to consider is the saturation point of the port with regard to valve curtain area versus port cross-sectional area. It's typically somewhere in the mid-lift range (roughly 0.300 to 0.400 inch) for most applications. Beyond this point, the valve curtain area becomes larger than the port cross-sectional area (c/s) and the port itself becomes the restriction. You can determine this point with the following formula:

Valve curtain vs. port saturation lift point =
valve lift x port c/s ÷ valve curtain area

Example: for a 2.02-inch valve at 0.400 lift and a port cross-sectional area of 2.15 square inches measured at the hump in the port wall adjacent to the pushrod.

Valve Curtain Area = 2.02 x 0.98 x 3.14 x 0.400 =
2.486 square inches

Port c/s = 1.87 x 1.15 = 2.15 square inches

Saturation Point = (0.400 x 2.15) ÷ 2.486 =
0.346-inch lift

So 0.346 inch lift is the point where the valve curtain area exactly equals the port cross-sectional area. Above this valve lift the port cross section becomes the controlling factor in flow capacity. (For an in-depth look at how much this affects cylinder head performance, take a look at *Chevy Small-Block Cylinder Heads* by Graham Hansen, published by CarTech.)

Calculating Optimum Port Area from Valve Size

The primary objective of all performance cylinder heads is to produce the maximum possible volumetric efficiency across the broadest possible range of engine speed. That's why port and valve sizes are so critical and so easy to mess up without careful deliberation. The intake port cross-sectional area (c/s) describes the smallest area of a port in a perpendicular plane to the flow upstream of the valve.

There are two schools of thought on this. Depending on the cylinder head, the smallest cross section may actually be the venturi diameter or throat area directly above the valve seat. This is particularly true if you also consider the additional obstruction of the valve guide and valve stem. Others define cross-sectional area as the choke point farther upstream near the bump in the port wall adjacent to the pushrod. To determine this you measure the vertical and horizontal dimensions at that point and multiply to find the area.

Port c/s Area = height x width

To find the area of the valve throat (venturi) simply measure the diameter of the throat opening above the valve seat and calculate the area as follows:

Throat Area = Pi x radius^2
Throat Area = diameter^2 x 0.7854

Head porters contend that the upstream cross-sectional area (in the port itself) should be 90 percent of the flow diameter of the intake valve for a race engine and 0.85 percent for a street engine. Some feel that 90 percent is good across the board. This is based on the valve's flow diameter at the inner edge of the valve seat. It's a reasonable assumption although the throat diameter directly above the valve seat may be even smaller and that is what the air actually sees. And it doesn't account for the partial blockage caused by the valve guide and stem. For now however, we are simply relating port cross-sectional area in the port itself to the flow diameter at the valve seat.

For example, a 2.02-inch intake valve has a flow diameter of 1.717 inches if we're going by the 85-percent rule. To calculate the equivalent port cross-sectional area, use the following formula.

Flow Diameter for Street Engine = valve diameter x 0.85

2.05 x 0.85 = 1.7425 inches

Port Area = (flow diameter^2 ÷ 4) x 3.1417

(1.7425^2 ÷ 4) x 3.1417 = 2.38 square inches

At 85-percent flow diameter ratio, the flow diameter is calling for a minimum cross-sectional area of 2.38 square inches in the port. To keep that in perspective, note that a 195-cc Air Flow Research cylinder head for a small-block Chevy has a 2.02-inch intake valve and a 2.21-square-inch cross-sectional area in the port. This is smaller than our calculation, but close. If we were to use the 90-percent rule it calls for a port c/s of 2.59 square inches, which is even larger.

It can be surmised that they're keeping it tight to preserve port velocity and they may also be considering the

Measure the valve flow diameter at the outer edge of the valve seat with dial calipers. If this is not possible, you can estimate it by subtracting 0.040 inch from overall diameter or by multiplying valve diameter by 0.98.

restriction of the valve stem. Up to a point, giving up CFM for port velocity is usually acceptable because velocity moves a fuel charge more effectively than area.

Calculating Minimum Port Area

Another way of evaluating port cross-sectional area comes from the (now out of print) SA-Design book *DeskTop Dynos* by Larry Atherton. His formula for calculating minimum port cross-sectional area offers an alternative method of estimating the minimum requirement based on cylinder volume times engine speed divided by an empirical constant of 190,000.

Minimum Port c/s Area =
(bore^2 x stroke x RPM) ÷ 190,000

A 350-ci engine with a 4-inch bore and a 3.48-inch stroke running at 6,000 rpm calculates as follows:

Area = (4.00^2 x 3.48 x 6,000) ÷ 190,000 =
1.758 square inches

That's almost dead on with a Chevy 492 casting, which has a cross-sectional area of 1.76 square inches. Many performance heads have larger cross-sectional port areas because they are trying to move as much air as possible while still maintaining port velocity. It is a delicate balance to strike and some do it better than others. It would be great to know the port velocity at the choke point, but it is rarely measured outside of an engine lab and calculating it would be difficult, particularly with a plenum involved upstream of the manifold runner. Atherton's formula provides a conservative but useful calculation of port area that lets you ballpark the minimum acceptable restriction based on engine size and RPM.

Calculating Port Velocity

If you know the cross-sectional area of a given port you can calculate the port velocity based on the bore diameter and the piston speed at any given RPM using the following formula.

$$\text{Port Velocity}_{fps} = (P_s \div 60) \times (B^2 \div A_p)$$

Where:
P_s = piston speed in feet per minute
B = bore diameter in inches
A_p = area of port in square inches

The first part of the formula converts the piston speed to feet per second while the second half relates the bore area to the port cross section. Consider the following example: A 3-inch-stroke 302-ci engine running at 4,400 rpm (torque peak) achieves a piston speed of 2,200 feet per minute at that point. The bore is 4.00 inches and the port cross section is 2.44 square inches.

$$P_{vel} = (2,200 \div 60) \times (4.00^2 \div 2.44) =$$
$$240.4 \text{ feet per second}$$

Estimating Peak Engine Speed from Airflow

SuperFlow Corporation (manufacturers of airflow benches, engine dynos, and chassis dynos) provides a handy formula for estimating the engine speed at peak power based on airflow and engine displacement. It uses empirically derived constants or VE factors to project peak-power RPM. Street engines use a VE factor of 1,196, while race engines use a larger factor of 1,316. A third

Measure the valve throat diameter (venturi) with a pair of dial calipers or a telescoping snap gauge. It should be close to 90 percent of the intake valve diameter.

To calculate port velocity, measure the port entry and exit and average the two area measurements to obtain A_p. Then plug in the mean piston speed and bore size to find the mean port velocity.

factor of 1,256 is also provided for more highly tuned street/strip applications.

It should be noted that these factors are only valid if you have airflow figures for a complete inlet tract at 28 inches of water. That means flowing the cylinder head with the intake manifold and carburetor attached to give the flow bench a look at the entire system. Say you have a 400-ci engine and you are able to obtain an appropriately measured flow value of 240 cfm through the complete inlet tract. Calculate the predicted peak engine speed based on the following formula and one of the three VE factors. For this example we'll use the street/strip factor.

VE Factors

1,196 = stock engine
1,256 = street/strip engine
1,316 = race engine

RPM = (VE factor ÷ displacement of 1 cylinder) x CFM

First divide 400 ci by 8 to get the displacement of one cylinder. It's 50 ci in this case.

RPM = (1,256 ÷ 50) x 240 = 6,028

(For more information about this formula and results from real-world applications, see *How to Build Max Performance Pontiac V-8s* by Jim Hand, published by CarTech.)

Predicting Horsepower from Airflow

Another SuperFlow-sourced formula along the same lines uses airflow through a complete system to predict peak horsepower. If you have already obtained the appropriate flow values and predicted your peak engine speed, you can use the same flow value to estimate peak power. I first learned of this simple formula from former SuperFlow Vice President Harold Bettes more than twenty years ago and have seen its prediction come pretty close many times. SuperFlow developed power coefficients for each of the most widely used flow bench test pressures. Again, remember that they are only accurate if you have airflow measurements through a complete inlet system.

Horsepower = observed CFM x power coefficient
x number of cylinders

Power Coefficient	For Flow at Inches of Water
0.43	10
0.35	15
0.27	25
0.26	28

Assuming a cylinder head with manifold and open carburetor attached, we observe a net flow of 225 cfm when tested on a flow bench at 28 inches of water. The corresponding power coefficient is 0.26.

Horsepower = 225 x 0.26 x 8 = 468

For best results you would want to verify flow numbers through several different ports.

Looking at Valve-Lift-to-Diameter Ratios

SuperFlow also pioneered a method for evaluating different valve diameters based on lift settings that are in direct ratio to their diameter. The lift-to-diameter (L/D) ratio provides ideal valve lift test settings that permit valve diameter flow comparisons based on common ratios. These ratios are 0.05:1, 0.10:1, 0.15:1, 0.20:1, 0.25:1, and 0.30:1. Ideal test valve lift settings are derived by multiplying the valve diameter by the desired L/D ratio. (See table on page 87.) To compare different size valves accurately, you first must select a common L/D ratio. A good choice would be one that yields a test valve lift that is about 65 percent of your total valve lift, since that is the point where the most flow occurs relative to valve open time.

Say you want to compare a 1.94-inch valve to a 2.02-inch valve, and you know that your total net lift is 0.450 inch. Multiply 0.450 times 0.65 to get the lift at 65 percent. That comes out to 0.292 lift. On the chart under the 1.94-inch intake valve column we find an L/D ratio of 0.15:1 that recommends a test lift setting of 0.291 inch. Note that the valve diameter of 1.94 inch multiplied by the L/D ratio (0.15) equals 0.291 inch.

To make an accurate flow comparison with a 2.02-inch valve, you use the same L/D ratio, which indicates 0.303-inch lift on the chart (2.02 x 0.15 = 0.303). Why more lift for the bigger valve? Because the flow increase is proportional. Testing at the same lift as the 1.94–inch valve would show a flow increase, but perhaps not as much as the 2.02-inch valve is actually capable of delivering.

That might lead you to think that it is not worth switching valves. Ask your flow technician to make the flow comparison using the appropriate L/D ratio to get an accurate picture of the valve's performance in proportion to the smaller valve. Then you truly know the

Lift-to-Diameter Ratios

Intake Valves			Exhaust Valves					
L/D	1.94	2.02	2.05	2.08	2.10	1.50	1.60	1.625

Test Valve Lifts

0.05	0.097	0.101	0.102	0.104	0.105	0.075	0.080	0.081
0.10	0.194	0.202	0.205	0.208	0.210	0.150	0.160	0.162
0.15	0.291	0.303	0.307	0.312	0.315	0.225	0.240	0.244
0.20	0.388	0.404	0.410	0.416	0.420	0.300	0.320	0.325
0.25	0.485	0.505	0.512	0.520	0.525	0.375	0.400	0.406
0.30	0.582	0.606	0.615	0.624	0.630	0.450	0.480	0.487

Courtesy SuperFlow Technologies Group

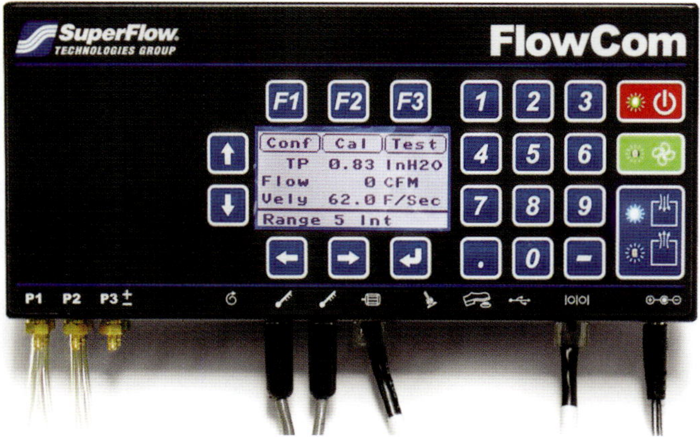

SuperFlow's FlowCom digital analyzer automates most of the data collection process. The software correlates and presents the collected data for easy analysis.

flow value of the larger valve and can make a more informed decision.

Comparing Flow at Different Test Pressures

SuperFlow is a primary manufacturer of airflow test benches and flow bench equipment. Their standard test pressure (water depression) is 25 inches, but most of the performance industry tests at 28 inches of water. For comparison purposes, SuperFlow developed a conversion chart for converting flow numbers at any test pressure to equivalent flow numbers at any desired test pressure.

The SuperFlow 600 is the most commonly used airflow bench in the performance world. It has the capacity, accuracy, and repeatability to test all modern performance cylinder heads.

Follow the chart below to select an appropriate conversion factor or do the math to calculate your own factor. If you have flow at a lower pressure (usually the case) and wish to convert to flow at a higher pressure, divide the higher pressure by the lower pressure and take the square root to obtain your conversion multiplier.

$$\text{Conversion Factor} = \sqrt{(\text{higher test pressure} \div \text{lower test pressure})}$$

You can verify this on the chart. Note that in some cases the chart uses a rounded number. For example, convert flow at 25 inches to flow at 28 inches.

$$\text{Conversion Factor} = \sqrt{(28 \div 25)} = 1.058$$

SuperFlow rounds this off to 1.06 on the chart. Now calculate an actual airflow conversion assuming 245-cfm flow through an intake port at 25 inches, and we want to convert it to equivalent flow at the performance industry standard of 28 inches.

$$\text{Flow at Known Test Pressure x Conversion Factor} = \text{flow at desired test pressure}$$

$$245 \text{ cfm x } 1.06 = 259.7 \text{ cfm at 28 inches}$$

The difference is minimal, so using rounded figures from the chart is perfectly acceptable. All the work has been done for you.

Now suppose you have 259-cfm flow taken at 28 inches and you want to convert it to flow at 25 inches. Just take the reciprocal of the known conversion factor. In this case:

$$\text{Reciprocal} = 1 \div 1.06 = 0.943$$

This is rounded to 0.945 on the chart. Multiply 259 cfm by 0.945 to get 245 cfm.

$$259 \text{ x } 0.945 = 244.75 \text{ (round to 245 cfm)}$$

In most cases you can rely on the chart for your conversion factor, although SuperFlow does stress that

Flow Bench Test Pressure Conversion Chart

Want flow at:

Have flow at:	3"	5"	8"	10"	12"	15"	18"	20"	25"	28"
3"	1.00	1.29	1.63	1.82	2.00	2.24	2.45	2.58	2.89	3.05
5"	0.774	1.00	1.26	1.41	1.55	1.73	1.90	2.00	2.24	2.37
8"	0.612	0.791	1.00	1.12	1.22	1.37	1.50	1.58	1.77	1.87
10"	0.548	0.707	0.894	1.00	1.09	1.22	1.34	1.41	1.58	1.67
12"	0.500	0.645	0.816	0.913	1.00	1.12	1.22	1.29	1.44	1.53
15"	0.447	0.577	0.730	0.816	0.894	1.00	1.09	1.15	1.29	1.37
18"	0.408	0.527	0.666	0.745	0.816	0.912	1.00	1.05	1.18	1.25
20"	0.387	0.500	0.632	0.707	0.774	0.866	0.949	1.00	1.12	1.18
25"	0.346	0.447	0.566	0.632	0.693	0.775	0.849	0.894	1.00	1.06
28"	0.327	0.422	0.534	0.598	0.654	0.732	0.802	0.845	0.945	1.00

To convert airflow measured at different test pressures, find your current test pressure in the first column. Then follow across the chart horizontally until you reach the column under the test pressure you wish to convert to. The number at that point is the multiplier for your conversion. For example, if you have a flow figure taken at a test pressure of 25 inches of water and you want to convert it to 28 inches of water, follow across on the 25 inch row to the last column under 28 inches. The number there is 1.06. Multiply your flow at 25 inches by 1.06 to obtain equivalent flow at 28 inches.

Courtesy SuperFlow Technologies Group

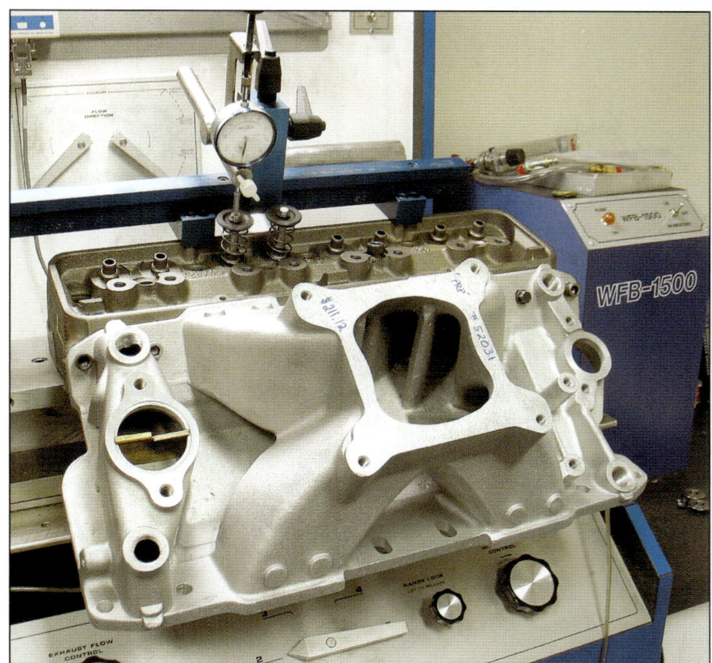

For even greater accuracy, it is often instructional to flow a cylinder head with the intake manifold attached. Some testers even install the carburetor with the throttle plates locked open to simulate the complete intake path.

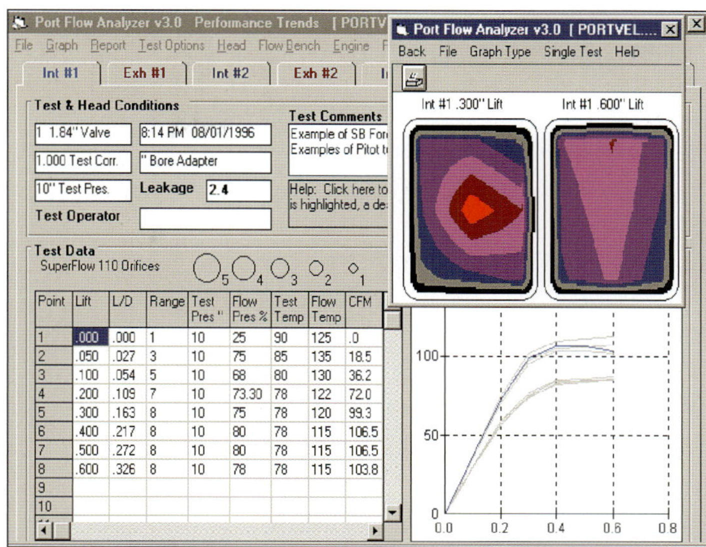

Performance Trends' Port Flow Analyzer Pro software allows you to match flow curves with camshaft profiles to create "flow area" curves that can help you evaluate a port's suitability for torque and horsepower potential. It is designed to interface with SuperFlow's Flowcom electronics and PT's automated valve opener to further computerize your flow bench testing.

greater accuracy is derived if the conversion range is smaller. So comparing 25 inches to 28 inches is more precise than comparing 10 inches to 25 inches. If you are testing on a SuperFlow flow bench equipped with the FlowCom digital analyzer, you can view these changes in the software. You can also make these conversions with Performance Trends Port Flow Analyzer software, which integrates easily with the FlowCom system.

Understanding Exhaust-to-Intake Ratio

The exhaust-to-intake (E/I) ratio is often misunderstood. Some still believe that it is the ratio of the exhaust valve diameter to the diameter of the intake valve, but we're not concerned with the ratio of valve size, which is largely dictated by available bore size. The E/I ratio is a percentage taken by dividing measured exhaust flow by intake flow at the same valve lift. This provides an index that can be used to compare cylinder heads based on flow efficiency.

E/I = exhaust flow at same lift ÷ intake flow at same lift

Example: 197 cfm @ 0.400 ÷ 247 cfm @ 0.400 = 80% E/I

Good heads usually fall somewhere in the 75- to 80-percent range, but this number typically varies according to lift. It may be 80 percent at mid-lift, but only 74 percent at peak lift as port efficiency diminishes. If you are able to obtain flow data on different heads, select those with the best E/I at the midpoint of your anticipated valve lift. While the valve sees peak lift only once per intake cycle, it sees mid-lift twice (in a sense) as it opens and closes. This is your point of greatest flow potential and it should be exploited with the best possible E/I ratio.

Most people don't ever consider many of the things discussed in this chapter, but by investigating them thoroughly prior to purchasing your heads, you ensure that you get the best possible package for your specific application.

If you wonder why the exhaust valve is always smaller than the intake valve, the answer is simple. On the intake side you're dealing with low pressure and relatively slow velocities. The exhaust valve operates under high temperature, high pressure, and more velocity. We have to coax the mixture into the cylinder on the intake side, but on the exhaust side it can't wait to get out.

Cylinder Head Formulas at a Glance

Combustion Chamber CID =
    measured cc ÷ 16.4

% Port Volume Difference =
    cc larger port ÷ cc smaller port x 100

Valve Curtain Area =
    valve diameter x 0.98 x 3.14 x valve lift
or (valve diameter − 0.040) x 3.14 x valve lift

Port c/s Area = height x width

Port Velocity$_{fps}$ = (P$_S$ ÷ 60) x (B^2 ÷ A$_p$)

    Where:
    P$_S$ = piston speed in feet per minute
    B = bore diameter in inches
    A$_p$ = area of port in square inches

Valve throat area = diameter^2 x 0.7854

Minimum Port c/s Area =
    (flow diameter^2 ÷ 4) x 3.1417

Minimum Port c/s Area =
    (bore^2 x stroke x RPM) ÷ 190,000

Valve Curtain ÷ Port Saturation Lift Point =
    (valve lift x port c/s) ÷ valve curtain area

Peak Power RPM =
    VE factor ÷ displacement of 1 cylinder x CFM

VE Factors
1196 = stock engine
1256 = street/strip engine
1316 = race engine

Estimated HP =
    CFM x power coefficient x number of cylinders

Power Coefficients
0.43 for flow at 10 inches
0.35 for flow at 15 inches
0.27 for flow at 25 inches
0.26 for flow at 28 inches

Lift-to-Diameter Ratio
Test Valve Lift = valve diameter x L/D ratio

L/D ratios
0.05:1
0.10:1
0.15:1
0.20:1
0.25:1
0.30:1

Airflow Conversion Factor (from lower to higher pressure)
    = higher test pressure ÷ lower test pressure

Airflow Conversion Factor (from higher to lower
    pressure) = reciprocal of conversion
    to higher pressure, or 1 ÷ x

E/I Ratio = exhaust flow at same lift ÷ intake flow at same
    lift

Valve Flow Diameter = valve diameter
    x 0.85 (street) or 0.90 (race)

Valve Flow Diameter for Calculating Valve
    Curtain Area = valve diameter x 0.98

Or

Valve Flow Diameter for Calculating Valve
    Curtain Area = valve diameter − 0.040 inch

Exhaust System Math

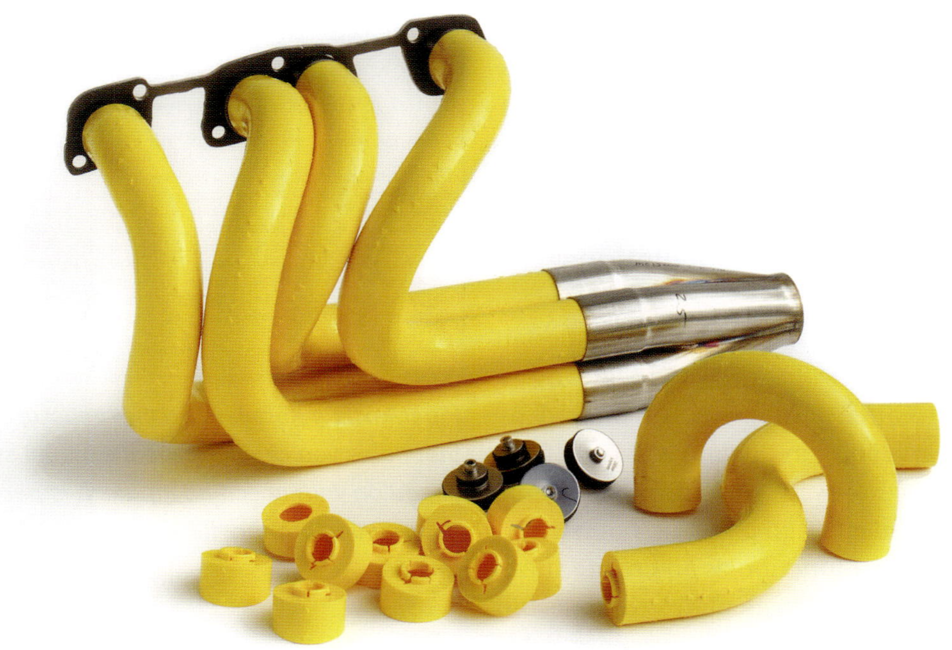

Exhaust systems don't generally invoke the glamour and gut level appeal associated with many high-performance induction systems, but they certainly play a critical role in the performance of every engine. Through decades of experience we have learned most of the ideal shapes, sizes, and configurations for high-performance exhaust systems, and we are able to calculate primary pipe sizes and lengths with a good deal of accuracy. It hasn't come easy, as the necessity for catalytic converters introduced detrimental restrictions that took a long time to overcome. Today we enjoy smaller, high-flow cat systems that are not the performance hindrance they once presented.

Of course, dual exhaust systems have long been recognized as a performance perk and interconnecting H-pipes and X-pipes have become the norm for high-performance exhaust systems. Three-inch and larger full-length exhaust systems have become prevalent and a broad range of appropriately sized high-performance mufflers accommodate them. So we have acquired a pretty good understanding of performance exhaust systems, but there are still ways to further optimize them through careful header selection based on engine displacement and RPM, volumetric efficiency, and final application. This is particularly true for applications unhindered by mufflers or those that run certain types of racing mufflers that have become more prevalent.

One drawback to the broad variety of header types and sizes is the potential for misapplication based on prevailing trends or which way the wind blows. Many enthusiasts still favor the "bigger is better" approach, which completely discards the purpose of building headers sized to suit specific applications. If the only purpose of headers were simply to look good, manufacturers would just build 2-inch Pro Stock–style headers for all small-blocks and 2½-inch headers for all big-blocks. Everyone would be happy, but performance would suffer dramatically.

In racing applications, exhaust systems are usually less restricted. Most often they involve a muffler requirement, stock exhaust manifolds or a possibly a spec header. Builders often consider mixed rocker arm selections optimized to provide appropriate exhaust event timing to aid cylinder blow down. Spec mufflers may be required and testing with back pressure readings can help pinpoint the most beneficial exhaust pipe cross-sectional area, effective primary pipe lengths, and collector sizes if headers are permitted. Modeling this on a PC simulator like PipeMax is a good idea if you want to pinpoint the best overall dimensions.

PipeMax Exhaust System Calculations

For those who want to get deeper into header design there is an inexpensive header-pipe-design program available on the internet called Pipe Max. It's offered by Meaux Racing Heads and you can download it from their website at www.maxracesoftware.com for a reasonable price. It is primarily a header design program, but it incorporates enough user data to function as an engine simulation program and it does calculate torque and horsepower output based on VE and the recommended header specs. It provides a wealth of good information about header dimensions and the proper design of headers to accommodate wave tuning. You input all the usual engine information along with details about your cylinder heads and camshaft and it calculates the optimum primary pipe diameters and cross-sectional areas, primary tube lengths and collector specifications for optimum performance. The results screen displays dimensions for a single primary pipe or 2- and 3-step header designs. It provides recommended pipe diameters and lengths for the primary pipes and the collectors. It also displays the best and worst specifications so you can avoid them. It calculates primary pipe harmonics for the first through eighth reflected wave and collector harmonics with mufflers and tailpipes. All of these specs are delivered in U.S. units and metric units automatically. The program is based on projected VE and it has a function to

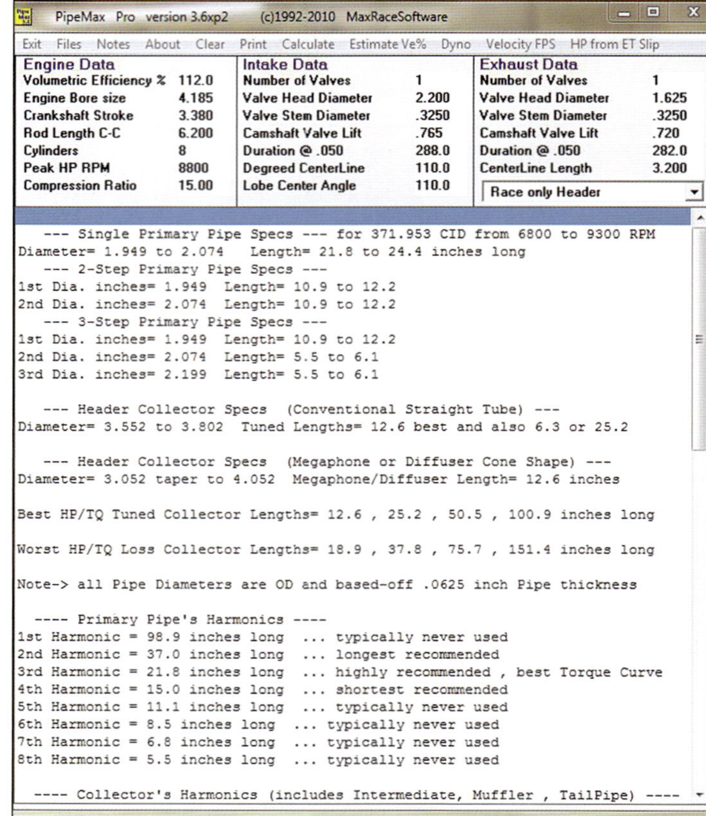

PipeMax is a popular and inexpensive software program that helps you calculate ideal exhaust system dimensions based on your specified inputs. It is particularly useful for determining primary tube lengths for step header configurations.

estimate VE for your particular engine specs. It allows you to select a dyno acceleration rate for the simulation and you can specify the mean flow velocity to be used in cross-sectional area calculations. While quite inexpensive, it is very thorough and makes a perfect complement to other simulators you might be using. It pinpoints the ideal header dimensions for your engine and simplifies header design. If you study it closely it teaches you a lot about the effects of pressure wave activity in the engine's exhaust system. I'm very comfortable in recommending it. Source: www.maxracesoftware.com.

Simulators are front loaded with all the mathematical fundamentals of engine performance. They can't always predict absolute power and torque, but they can illuminate trends according to the physics and that amounts to a pretty good road map for serious engine builders. As long as the requirement is clearly identified in the rules,

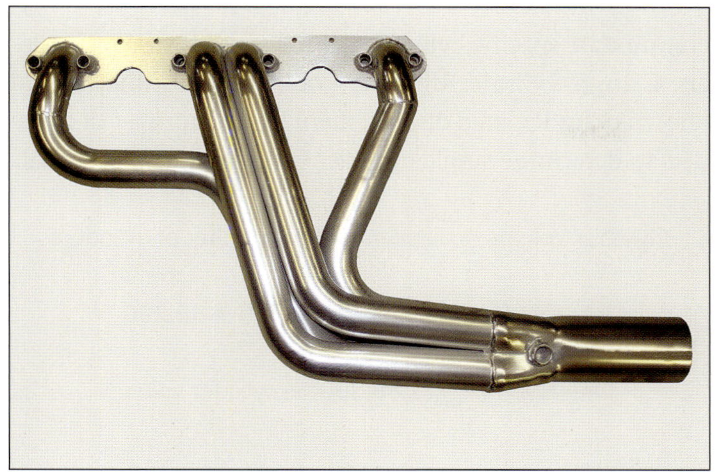

Equal length 4-into-1–style headers consistently outperform all other types, particularly when primary pipe cross-sectional area is well matched to engine displacement and VE at torque peak RPM.

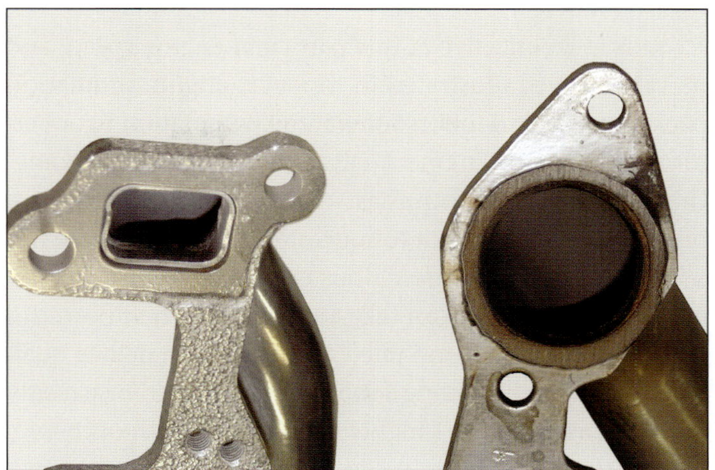

Note the difference in cross-sectional area (c/s) of these two headers. The formula for primary pipe c/s area addresses the optimum cross section based on cylinder volume and engine speed at the torque peak.

savvy engine builders always find ways to optimize within those rules. If a spec header is involved and you're not comfortable with discreetly modifying pipe dimensions, consider the possibility of different cam timing in cylinders whose intake and exhaust timing peaks differ due to unequal port or pipe cross section or length. Like other components, any restriction on the exhaust system should be carefully evaluated for its effect on torque production and how it affects other parts of the engine package. There are usually ways to refine and assemble a more robust package even within restrictive rules.

I once did a dyno test with a client whose big-block Chevy arrived with 2¼-inch headers, oval port cylinder heads, a conservative 218-degree camshaft, and a dual-plane intake manifold sporting an 850-cfm Holley 4-barrel. When it didn't perform as expected, he was thoroughly disappointed and immediately surmised that something was wrong with the dyno. I suggested trying a set of 1¾-inch headers that I had in my testing inventory, but he insisted that his high-performance big-block needed the big-tube headers that he had seen on everyone else's car at the track. The fact that he had a street engine and that everyone else was running big-tube headers regardless of application simply reinforced the depth of the problem. I finally convinced him to try the smaller headers, and he was astonished when they delivered a solid 38–ft-lbs gain at the torque peak and raised the power curve by about 20 hp across the board.

I knew from personal experience that this was the right move and any other good dyno shop would have done the same thing. The point is simple: Just because the catalog lists big-tube headers doesn't necessarily mean that you need them to achieve the best performance from your particular application.

Calculating Primary Tube Cross Section

So, how do you determine the correct header size for any given engine combination? Copying the other guy amounts to a shot in the dark and trial and error is pretty expensive. Wave tuning theory has long been a popular method, but it is typically beyond the reach of the average enthusiast. Moreover, it can be complicated and is probably less effective when saddled with full-length exhaust systems that include catalytic converters, X-pipes, supersonic master mufflers, and all the latest "gotta have it" exhaust goodies. Most of these pieces serve a legitimate function, but their presence complicates wave tuning.

In short, wave tuning seeks to take advantage of oscillating pressure pulses within the exhaust stream to aid cylinder scavenging. The theory is to time the pulses so that a reflecting pulse arrives at the cylinder at just the right time to lend its energy to carrying exhaust gases out of the cylinder. The timing is a function of primary tube length, and it is affected by the point where the pulse reaches atmospheric pressure at the end of the tube.

It is most effective at one particular engine speed, and its value is well illustrated in top-level engine simulation packages such as Motion Software's Dynomation 5, or Performance Trend's Engine Analyzer Pro. But it is not something the average enthusiast can jump right into on a one-off basis. This is not to suggest that simulators are not a valuable tool for more serious enthusiasts with real racing applications that can take advantage of their powerful capabilities. Both are top-rated simulators that deliver quality results while providing a solid learning experience neatly disguised as pure fun. I highly recommend both of them.

For the desktop engine designer only equipped with a catalog and a hand-held calculator, there is still a way of calculating primary header pipe diameter for effective performance based on engine displacement, engine speed, and volumetric efficiency. In the interest of raising awareness and furthering everyone's performance goals, leading performance authority Jim McFarland has provided a simple formula for calculating the optimum cross-sectional area ($A_{C/S}$) of a header primary tube or pipe.

This method optimizes for the engine's torque peak or point of maximum volumetric efficiency and thus the point of maximum exhaust volume. At engine speeds above the torque peak, cylinder filling decays proportionately with RPM (time) because there is progressively less time available to fill the cylinder on each intake event. There are more power strokes per minute, but proportionately less exhaust volume to evacuate due to declining VE. Optimizing for the torque peak provides the ideal primary pipe diameter, and since peak power is usually no more than 1,500 to 1,750 rpm above the torque peak, the selected pipe size has no trouble accommodating an exhaust volume that essentially flatlines, and then begins to fade rapidly.

The RPM spread between the torque peak and the power peak is influenced by the engine's rod-to-stroke (R/S) ratio (see Chapter 2). Given a constant stroke length, a shorter rod tends to broaden the separation between the torque peak and the power peak. Similarly, a longer rod tends to move the peaks closer together. The previously stated range applies for most of the stroke and rod lengths commonly used by performance enthusiasts. The formula can be written two ways: one to solve for primary pipe c/s area when the torque peak RPM is known or anticipated, and the other to predict torque peak engine speed based on a primary pipe cross section that is being considered. To ac-

commodate specific cylinder volumes, the formula considers a single cylinder only, so you'll have to divide your known or anticipated displacement by the number of cylinders to determine single cylinder volume or displacement.

$$\text{Cylinder Volume} = \text{displacement} \div \text{number of cylinders}$$

$$A_{c/s} = (\text{cylinder volume} \times \text{RPM}) \div 88{,}200$$

$$\text{Or RPM} = (A_{c/s} \times 88{,}200) \div \text{cylinder volume}$$

Where:
$A_{c/s}$ = primary pipe c/s area
Cylinder volume = volume of a single cylinder
88,200 = mathematical constant
RPM = RPM at torque peak

Once we know the calculated c/s area we can calculate the corresponding primary pipe diameter to the nearest available pipe size.

$$\text{If Area} = \text{diameter}^2 \times 0.7854$$

Then pipe size equals the square root of the previously calculated c/s area times the reciprocal of the constant 0.7854.

$$\text{Pipe Size} = \sqrt{[A \times (1/0.7854)]}$$

$$\text{Or}$$

$$\text{Pipe Size} = \sqrt{(A \times 1.273)}$$

If you're solving for a torque peak RPM based on contemplated primary pipe size, calculate the c/s area by squaring the inside diameter (ID) of the pipe and multiplying by 0.7854. To determine the true inside diameter of a pipe for the purpose of calculating c/s area, use the measured outside diameter (OD) minus twice the wall thickness.

$$\text{ID} = \text{OD} - (2 \times \text{wall thickness})$$

Example:
1.75-inch OD pipe with a wall thickness of 0.040 inch

$$\text{ID} = 1.75 - (2 \times 0.040) = 1.67 \text{ inches}$$

As performance exhaust systems encounter various restrictions such as catalytic converters and some types of high-performance mufflers, peak power can be eroded. Accordingly, header pipe size selection based on manipulating peak torque RPM points outweighs the pursuit of absolute peak power. This suggests erring on the small side of primary pipe selection to preserve velocity with the minimum c/s required to service cylinder volume at peak torque; hence the formula for c/s area based on engine speed, cylinder volume, and VE. Additionally, it should be noted that tuned induction systems generate their own independent torque curves that contribute proportionately to an engine's "net" torque peak RPM.

The default header size for most small-block engines seems to be 1¾, or 1.75, inch. Whether or not this is really the best choice depends on individual cylinder displacement and the exhaust volume it generates at peak torque.

The accompanying chart shows some calculated examples based on popular muscle car engines. Note the advertised torque and the RPM at which it occurs, the individual cylinder volume, calculated c/s area, calculated pipe size, and closest available pipe size. Note further that all of these performance engines except perhaps the larger-displacement L99 Camaro and the earlier 350 SS call for a primary pipe diameter smaller than 1¾ inch, yet it is extraordinarily difficult to convince people to run smaller appropriately-sized headers on street cars. All major header manufacturers offer correctly-sized headers for these cars, but many enthusiasts ignore the recommended size and choose the next larger size because their gut tells them it is necessary. In most cases, it's not.

Referring back to the previously cited production engines, we note that the 302-ci Z28 calls for a 1.79-square-inch c/s and 1.51-inch ID primary pipe to complement its exhaust volume at the 4,200-rpm torque peak. Consulting the chart, we check the c/s area and corresponding ID's for a header with 16-gauge primary pipes (most common). The closest match is a 1.625-inch OD pipe that offers a 1.507-inch ID with 1.783-square-inch c/s area. In a catalog that would be listed as a 1⅝-inch header with 16-gauge primary pipes and it is a near perfect match for the 302 engine.

Some heavy-duty headers use 14-gauge primary pipes. In this case you should still select the 1⅝-inch (1.625) primary pipes even though the c/s area is a bit smaller. The next-larger common pipe size is 1.750, but the 14-gauge c/s area of 2.010 inches would be excessive for the stock 302's exhaust volume. It would tend to reduce torque and shift the torque peak to a higher RPM. Also recognize that even though you might rev your 302 to 6,500 rpm, VE is proportionately reduced at that engine speed and the declining exhaust volume will not overtax the indicated primary pipe. Again, the key to primary pipe selection is the inside (ID) diameter c/s area indicated for the engine's torque peak.

To cite a further example, we can look at the results of the 2009 Engine Master's Challenge dyno shootout produced by *Popular Hot Rodding* magazine. Four-time winner John Kaase captured first place with a 403-ci small-block Ford that delivered 677 hp at 6,400 rpm and 597 ft-lbs of torque at 5,400 rpm through the mufflers. He ran Hedman headers with 2-inch-diameter primaries. Now let's see how well the formula matches up to this selection.

Muscle Car Calculated Cross-Sections and Primary Pipe Diameters

ENGINE/ci	TORQUE @ RPM	CYLINDER VOLUME	CALCULATED $A_{c/s}$ (D^2 X 0.7854)	CALCULATED ID ($A_{c/s}$ X 1.273)	CLOSEST OD PIPE DIAMETER
1969 Z/28 302	290 @ 4,200	37.75 ci	1.79 sq. in.	1.51 in.	1⅝ (1.625) in.
1987 IROC 305	290 @ 3,200	38.12 ci	1.38 sq. in.	1.32 in.	1½ (1.500) in.
1968 Camaro SS 350	380 @ 4,200	43.75 ci	2.08 sq. in.	1.63 in.	1¾ (1.750) in.
2010 Camaro 376	420 @ 4,300	47.00 ci	2.45 sq. in.	1.76 in.	1⅞ (1.875) in.
2010 Dodge Hemi	410 @ 4,300	43.12 ci	2.10 sq. in.	1.63 in.	1½ (1.500) in.
2010 Mustang 281	325 @ 4,250	35.12 ci	1.69 sq. in.	1.46 in.	1½ (1.500) in.

Primary Pipe Selection Guide

ID = OD – (2 x wall thickness)

| 14 gauge = 0.075 inch | | | 16 gauge = 0.0625 inch | | | 18 gauge = 0.047 inch | | |
OD	ID	c/s Area	OD	ID	c/s Area	OD	ID	c/s Area
1.500	1.350	1.431	1.500	1.382	1.500	1.500	1.406	1.552
1.625	1.475	1.708	1.625	1.507	1.783	1.625	1.531	1.840
1.750	1.600	2.010	1.750	1.632	2.091	1.750	1.656	2.153
1.875	1.725	2.337	1.875	1.757	2.424	1.875	1.781	2.491
2.000	1.850	2.688	2.000	1.882	2.781	2.000	1.906	2.853
2.060	1.910	2.865	2.062	1.942	2.962	2.062	1.966	3.035
2.125	1.975	3.063	2.125	2.007	3.163	2.125	2.031	3.239
2.250	2.100	3.463	2.250	2.132	3.569	2.250	2.156	3.650
2.375	2.225	3.888	2.375	2.257	4.000	2.375	2.281	4.086
2.500	2.350	4.337	2.500	2.382	4.456	2.500	2.406	4.546

403 ci ÷ 8 = 50.37 (rounded to 50.4 cid per cylinder)

$$A_{c/s} = 50.4 \times 5{,}400 \div 88{,}200 = 3.085 \text{ square inches}$$

Therefore:

$$\text{Primary Pipe Size} = \sqrt{3.085 \times 1.273} = 1.98\text{-inch ID}$$

Pretty close, with minor accommodation for tube thickness. Note that the object of the Engine Master's Challenge is to maximize torque, and thereby horsepower, across the broadest possible range of engine speed. An experienced engine builder like Kaase leaves nothing on the table, and our formula neatly confirms his chosen header size. Moreover, he entered a second engine in the challenge, a 511-ci monster based on a 429 Ford. It happily thumped out 856 hp at 6,400 rpm and 751 ft-lbs of torque at 5,500 rpm. Again he selected a Hedman header, but this set had 2.25-inch primaries.

$$511 \div 8 = 63.875 \text{ ci per cylinder}$$

$$A_{c/s} = 63.875 \times 5{,}500 \div 88{,}200 = 3.983 \text{ square inches}$$

Therefore:

$$\text{Primary pipe size} = \sqrt{3.983 \times 1.273} = 2.251 \text{ inches}$$

Right on the money. Note that a key component of this formula is individual cylinder volume and the number of times you have to process (evacuate) it per minute. Because the torque peak represents maximum VE, we know that it is the point of greatest cylinder filling, highest charge density, and maximum exhaust volume. By optimizing header capability to match peak VE we achieve optimum system resonance at the point of maximum efficiency. Max torque follows accordingly.

My thanks to Jim McFarland for providing this simple formula and kudos to Mr. Kaase for proving it so handily. Because it addresses exhaust volume at peak VE, it calculates the optimum primary pipe diameter independently of primary tube length. The calculated area delivers optimum exhaust efficiency because it is ideally sized to move exhaust volume based on cylinder displacement and processing speed (RPM). Within reason, the calculated pipe diameter will deliver top performance regardless of length as long as the individual lengths can also be juggled to help crutch deficient inlet flow paths as found on good-port/bad-port big-block Chevys.

Once a peak torque RPM has been established, changing primary pipe length will not change the RPM point, but it will cause torque values to pivot back and forth about that point. Lengthening the primary pipe tends to inflate the torque curve below peak torque, while shortening the primary pipe typically shifts torque value from below the peak and adds it above the peak. Again, it is the change in primary pipe diameter and thus c/s area that moves the torque peak up or down the RPM scale.

For your convenience, the Primary Pipe Selection Guide (see page 96) is a list of commonly used primary pipe diameters and their corresponding c/s areas based on available pipe diameters and wall thickness. Once you calculate your ideal c/s area and corresponding ID, locate them on this chart to determine the most appropriate primary pipe size for your header selection.

Calculating Primary Tube Length

Despite the previous discussion, there is a formula for calculating a preferred primary tube length. In his book, *Performance Tuning in Theory and Practice*, A. Graham Bell cites empirical formulas for primary pipe length and diameter that yield ballpark results surprisingly close to the previously examined exhaust volume and VE formula.

$$L_{in.} = [850 \, (360 - EVO) \div rpm] - 3$$

$$Dia._{in.} = [(vol. \times 16.38) \div (L + 3)25] \times 2.1$$

Where:
L = primary pipe length
EVO = exhaust valve opening point from cam card or internet
vol. = volume of a single cylinder
Dia. = calculated primary tube diameter

Using these formulas with the same information from John Kaase's 403-ci Engine Master's Challenge engine, we calculate a primary pipe length of 39.8 inches and a primary tube diameter of 2.06 inches. Since we don't know his cam specs other than 246 degrees at 0.050-inch lift and 0.750-inch total lift, we can speculate a pretty racy camshaft with EVO at 88 degrees BBDC. If we're wrong and we drop it down to 77 degrees BBDC, we calculate a primary pipe length of 41 inches. The former is more likely correct.

Calculating Collector Diameter and Length

Generally speaking, header collectors do their best work at or below peak torque RPM. Adding collector vol-

ume typically increases torque in this range while reducing collector volume tends to diminish torque values to a degree. In discussions with Jim McFarland, he pointed out the sprint car trend of running headers with virtually no collectors to limit low-end torque, which can be detrimental on dirt tracks. Similarly, Pro Stock headers merely merge the primary tubes and don't use a collector since they never see operation below peak torque.

Collector dimensions are, at best, difficult to pinpoint and there are no reliable formulas to accommodate all the variables, including large area changes and the amplitudes of reflected waves as they enter the collector. Motion Software's Larry Atherton suggests that smaller collectors offer less area change, thus reducing the amplitude of the first reflected wave and encouraging mid-range power. Larger collectors generate stronger waves at the primary/collector interface and since this is closer to the cylinder it tends to increase top-end power. For the ballpark crowd, Atherton suggests the following formulas for estimating collector length and diameter:

$$Collector \, Diameter = 1.9 \times primary \, pipe \, diameter$$

$$Collector \, Length = 0.5 \times primary \, pipe \, length$$

These are ballpark formulas, but they work pretty well. We previously calculated the primary pipe diameter of John Kaase's 403-ci Ford at 2.00 inches. Using this formula we calculate 1.9 x 2 = 3.8-inch diameter for his collectors. We also calculate a primary pipe length of 39.8 inches, which yields a 20-inch collector according to our formula.

$$Collector \, Diameter = 1.9 \times 2 = 3.8 \, inches$$

$$Collector \, Length = 0.5 \times 38.9 = 19.9 \, inches$$

In practice it appears that Kaase used a slightly smaller merge collector fitted to performance mufflers, so we can't be certain. Still, you can follow this example and use the formulas to select collectors based on your particular input.

Food for Thought

Another point relative to primary pipe section area and peak torque RPM suggests the possibility of further

Header Modeling and Design Kit

Icengineworks offers a header design kit for header shops, race shops and hardcore enthusiasts who want to build their own headers to fit a particular chassis requirement. This clever kit is like nothing you've seen before in that it provides a complete set of ABS plastic modeling blocks shaped exactly like header tubing of the appropriate size. Each block is exactly one inch long and they come in straight and curved versions that provide two-, three-, four- and six-inch radius bends at the centerline. You can snap these blocks together in any combination and rotate the curved pieces to achieve any desired configuration and radius. It allows you to piece together a full-scale model header that exactly fits your engine and chassis combination. Each individual piece is clearly marked with multiple indexing arrows andwitness marks so you can easily duplicate the sections in metal tubing using appropriately cut lengths of straight, J-bend, or U-bend tubing. The witness marks are spaced 30 degrees apart, providing exact positioning references for transferring the design to metal. The Pro Kit version includes expandable adapters that lock the blocks in header flange starter tubes so the initial blocks will not move during the process. The beauty of this system is that it allows you infinite freedom to rotate, extend, or shorten any segment into the desired shape and length you need. This makes it particularly easy to model your way around various chassis obstructions and still achieve an equal length header or a multi-length header with perfectly aligned joints and bends. It gets even better from there. The company also offers construction aids in the form of a special pivot table for making precise tubing cuts, starter block accessories, and special tack welding clamps that facilitate very precise joints for final welding. You can purchase a kit for just one side or complete kits designed to model both sides of a V8 engine. Once you have created a perfect model header and transferred it to metal, you can break down the model and store the kit until the next header project comes along. The math involved is easy and straightforward, consisting primarily of counting the segments and noting the appropriate lengths and radii. It couldn't be easier and it can turn an amateur into a first-class header designer almost overnight. Source: www.icengineworks.com.

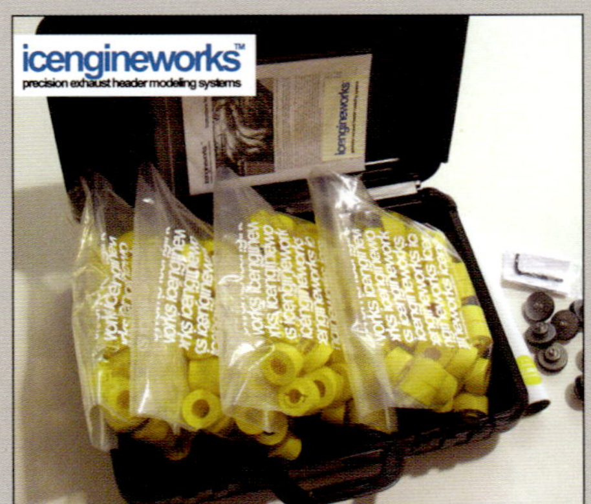

Icengineworks kits come in a handy ABS case for storage. The blocks snap together and come apart with minimal force so they're very easy to use. This kit will model one side of a V-8 engine with 2-inch diameter pipes (yellow).

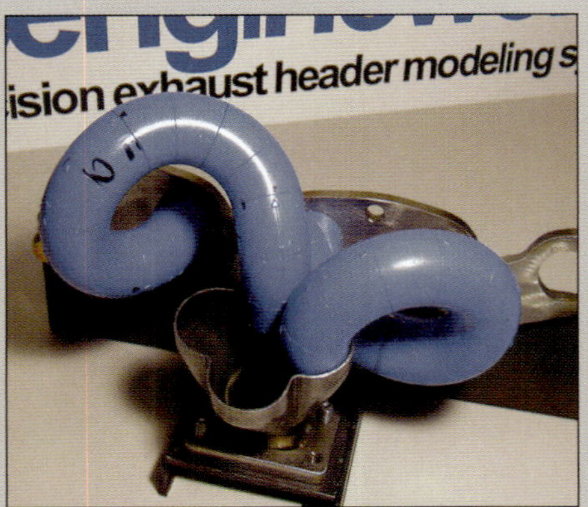

Icengineworks unique header modeling blocks rotate and position to form even the most complex header requirements such as this partially completed turbo exhaust manifold. Note the identical 1-inch curved sections and how they easily rotate to any desired position.

The pivoting table and spacer work on almost any vertical band saw; providing perfect cuts that are always perpendicular to the centerline of the metal tubing so it is very easy to duplicate the design of the modeled header in metal.

Here a complete header design has been modeled on a V-8 engine already installed in a chassis. Note the starter blocks and handwritten notes on each segment indicating the lengths and degree of alignment so the header fabricator can easily duplicate the finished model in metal.

The design is converted to metal with the tack welding clamps in place and perfectly aligned joints. If you're not a competent welder you can duplicate the model in this fashion and take it to a welding shop for final welding. Note how the kit allowed the designer to create perfectly matched free-flowing bends.

The pivot table, tack welding clamp kits and other accessories must be purchased separately, but they function similarly with all of the modeling block kits. Once you have all the components of the system you can model and build any type of header required.

This demonstration jig shows the starter blocks as they are attached to a simulated head. The metal tubing is measured and cut from the precise modeling block dimensions to duplicate the exact header as previously modeled on the engine and chassis.

Cleverly designed clamps provide perfect pipe alignment for tack welding. By anchoring both sections securely, they create a gapless, easy-to-weld joint with smooth connections and no sharp bends or abrupt changes in direction.

broadening a torque curve by constructing headers with two different primary pipe diameters alternating in size according to firing order. Jim McFarland patented a header and intake manifold system using just such an arrangement. By arranging the smaller intake runners to correspond with the smaller c/s area primary pipes and similarly for the larger runners and primary pipes, it is possible to effect two torque peaks in the net curve, thus further exploiting the relationship between c/s area and peak volumetric efficiency (torque).

Exhaust System Wave Dynamics

It may be useful to some readers to contemplate the function of wave dynamics within an exhaust system. This brief explanation barely scratches the surface of a complex discipline, but its intent is to illuminate the prevailing thought as it applies to the action of finite amplitude waves operating within the exhaust flow.

Finite amplitude waves are invisible pressure waves moving back and forth through an exhaust primary pipe completely independent of the flow of exhaust gas particles. There are two types: compression waves and expansion waves.

Compression waves are positive pressure disturbances with a pressure ratio greater than one relevant to ambient pressure within the primary exhaust pipe. Expansion waves have a strong negative pressure (less than 1).

These waves are a physical phenomenon traveling much faster than the gas particles in the exhaust stream. Because the exhaust system operates at much higher pressures than the induction system, these waves (or pulses) can be very strong, and they can exert considerable influence on the flow of exhaust gases. Positive pressure "compression" waves help push gas particles in the same direction that the wave is traveling, while negative pressure "expansion waves" help push gases in the opposite direction of their travel. This means that both waves can assist in the evacuation of exhaust gases from the cylinder.

Near the completion of a combustion event, the exhaust valve opens and a compression wave moves through the exhaust port into the header primary pipe. This high-pressure wave lends its energy to the outgoing gas flow and drives it toward the end of the pipe, which may or may not end at a header collector. It is the positive

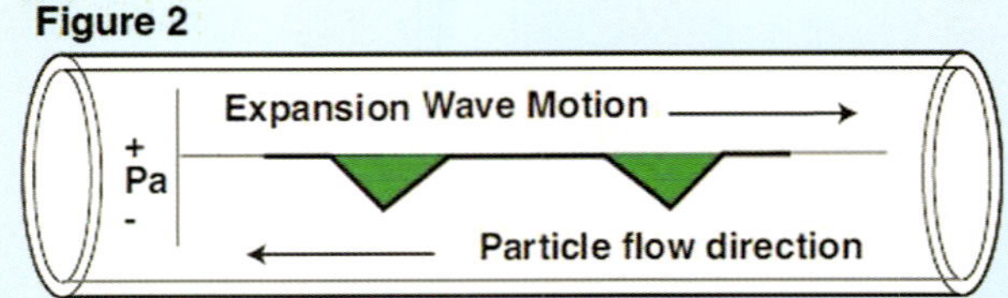

Gas flow in both of the primary pipes shown here is moving left to right, so the cylinder is off to the left and the primary pipe ends to the right. Note the plus and minus symbols indicating the compression wave is positive and the expansion wave is negative. The waves move in opposite directions, but both apply energy to moving gas particles toward the header pipe exit. Shorter pipes return a scavenging wave sooner and achieve "tune" at higher engine speeds. Longer pipes delay arrival of the scavenging wave and tune to lower engine speeds. (Courtesy Larry Atherton, Dynomation 5 manual)

wave moving gas particles in the same direction of travel. When it reaches the end of the pipe, a negative pressure expansion wave of near equal intensity is reflected back up the pipe toward the cylinder. Since negative pressure waves push gases in the opposite direction of wave travel, the expansion wave also helps move gases toward the primary pipe exit. When this wave reaches the cylinder, it arrives with a significant drop in pressure. If, through careful tuning of the primary pipe length, it can be timed to arrive at the beginning of the valve overlap period (both valves slightly open at the same time), the low pressure will create a pressure differential (vacuum) in the cylinder which allows atmospheric pressure to push additional air and fuel into the cylinder.

This effect is called scavenging and it accomplishes three things. It assists the flow of outbound exhaust gases, it tugs on the incoming fuel charge to help initiate cylinder filling, and it helps purge the cylinder of residual exhaust gases. The timing of these events is critical and is largely dictated by engine speed and length of the primary pipe.

The timing and intensity of wave action is controlled by primary pipe length and diameter. In the application of wave dynamics, primary pipe size is reflective of overall cylinder volume and the relationship of pipe size to the intensity of pressure waves. Large-diameter pipes generate lower pressures. Correspondingly lower positive pressure waves reflect lower amplitude suction waves that are less effective at scavenging residual exhaust and encouraging cylinder filling. Smaller pipes create higher pressures, which generate stronger scavenging waves but increase restriction and pumping effort. The correct balance can be effective depending on cylinder displacement (and corresponding exhaust volume), engine speed, and camshaft timing with regard to exhaust valve opening point and overlap period.

Primary pipe length dictates the timing of the suction wave's arrival at the cylinder. With shorter tubes the scavenging wave arrives sooner and achieves tuning resonance at a higher RPM. The tradeoff is reduced cylinder blowdown due to the decrease in available time. This retains higher cylinder pressure and increases pumping work against the piston trying to empty the cylinder. Opening the exhaust valve earlier helps, but it forfeits cylinder pressure that could still be driving the piston. With early exhaust valve opening, larger pipes are required to optimize scavenging wave arrival time.

Since these wave changes occur many times during a single exhaust event, they also become progressively weaker. There are counteracting forces at work here. Higher engine speeds require shorter pipes and earlier exhaust valve opening, but early opening requires longer pipes to optimize scavenging. A delicate balance is required.

The addition of collectors on headers further complicates the matter, but with potentially beneficial results. With no collector, the individual tubes create stronger suction waves that tend to peak within a narrow range of engine speed. Collectors improve performance and broaden the effective range by extending the width of the returning suction wave. Larger diameter collectors present a more direct path to atmospheric pressure and strengthen the front of the suction wave while smaller collectors tend to strengthen the end of the suction wave. The length of the collector controls the timing between the leading and trailing edge of the suction wave and can thus affect overall tuning across all of the pipes connected to it.

Because there are overlapping tradeoffs to wave tuning, other factors come into play. Equal-length primary pipes are well known to boost power by equalizing the work performed by each cylinder. But according to wave dynamics, they function best at or near one optimum speed depending on their length. Equal lengths often require more bends, which may increase restriction in some pipes, but wave dynamics infers that unequal-length tubes with minimal bending can effectively broaden the power range.

Of course this does not account for the presence of mufflers, catalytic converters, and convoluted tailpipes. Hence it more likely applies to racing applications where these elements are not present. Most race engine builders do not agree with this theory, but it further suggests that wave tuning action might be useful in crutching "bad" intake ports like those found on big-block Chevys. If the "poor" ports can be wave tuned (via the exhaust) to match the "good" ports, overall performance improves. There may be more of this type of tuning going on than is readily apparent, and whatever success it might provide is, for the most part, unreported because nobody's talking.

Due to the complexity of tuning elements involved, wave dynamics is probably best left to professional racing efforts that are equipped and funded to pursue it. Still, the subject is fascinating, and armchair tuners willing to go the extra mile can study and apply it through the miracle of PC simulation. Motion Software's Dynomation 5 Wave Action Simulator and Comp Cams' DeskTop Dyno 5 are highly effective simulators that can lead the casual tuner to a higher level of expertise with regard to wave action tuning. Performance Trends' Engine Analyzer Pro also incorporates wave tuning simulation to predict engine performance. If you're a street enthusiast who just wants to make a good decision on a set of off-the-shelf headers, you should focus on the relationship between cylinder volume and torque peak RPM to choose your primary pipe size. But once you get a feel for wave action simulators you'll really enjoy working with them.

Header Tips

- Changing the length of header primary pipes does not change the RPM where peak torque occurs. It does tend to fatten or diminish the torque curve slightly on either side of peak torque depending on the direction and magnitude of change and the influence it exerts on wave dynamics within the pipe.
- Peak torque RPM can be influenced up or down by increasing or decreasing primary pipe diameter. To whatever degree the exhaust header contributes to torque, a larger pipe extends the torque peak higher in the RPM range while a smaller pipe tends to shift the torque peak toward a lower RPM depending on the net change in cross-sectional area (c/s) and the degree that it affects mean flow velocity in the pipe.
- Headers with stepped primary pipes can broaden an engine's torque curve by providing additional degrees of freedom volume that encourages multiple torque boosts based on the desired mean flow velocity.
- Header collectors work best at or below peak torque RPM and are of little value above it. Because peak torque represents peak exhaust volume, a collector that adequately serves the engine's needs is amply sized to carry it through to peak power. Many applications actually forgo a collector altogether if the engine operates above peak torque most of the time.
- Crossover pipes in a dual-exhaust system tend to increase torque below the torque peak by effectively adding collector volume.
- You can broaden a torque curve by building a header system with two different-sized primary pipes arranged alternately in the firing order. They will tend to establish two separate torque peaks at different engine speeds, which effectively fattens the curve.
- Headers and intake systems can be optimized separately to produce two distinct torque peaks that also broaden a given torque curve.
- Headers sized to promote torque peak optimization reduce back pressure and improve combustion efficiency through faster flame travel. This permits a reduction in initial timing and reduces work opposing the piston as it approaches TDC.

Exhaust System Formulas at a Glance

Primary Pipe c/s Area = volume of 1 cylinder x RPM at peak torque ÷ 88,200

Primary Tube Diameter = c/s area x 1.273

Torque Peak RPM = primary pipe c/s area x 88,200 ÷ volume of 1 cylinder

Primary Tube ID = measured OD – (2 x wall thickness)

Primary Pipe Length = [850 (360 – EVO)] ÷ RPM – 3

    where EVO = exhaust valve opening point from cam card

Primary Pipe Diameter = [(volume of 1 cylinder x 16.38) ÷ (calculated length +3) 25] x 2.1

Collector Diameter = 1.9 x primary pipe diameter

Collector Length = 0.5 x primary pipe length

FUEL SYSTEM MATH

To calculate and service the requirements of your engine's fuel system, you need a basic understanding of brake specific fuel consumption (BSFC), fuel flow requirements, injector sizing, jet area calculations, pump sizing, and other factors that affect how your engine burns fuel and how to optimize your fuel system. You'll find that you use the formulas in this chapter more frequently than the primary engine building formulas because the information presented here affects how you operate and tune the high-performance engine you strived so diligently to construct.

Understanding BSFC

Brake specific fuel consumption (BSFC) is frequently misunderstood. Many people mistakenly believe that it is an indicator of rich or lean fuel mixtures, but it is actually a measure of efficiency that indicates how well the engine uses the fuel it burns. More specifically, it is the rate in pounds of fuel per horsepower per hour that a given engine consumes to make power. There is a range of optimum efficiency for most engines, and BSFC defines that range.

As you may have already surmised, the term "brake" precedes it because BSFC is usually measured with an engine running on a dyno. BSFC figures are typically quoted for wide-open-throttle conditions, but it is also a measurable quantity that relates to fuel economy at part throttle operation. In the performance world, we use it to judge the efficiency contribution of various engine combinations and to predict certain requirements such as fuel injector flow rate.

A particular cylinder head may be found to make more power with less fuel, and that's an indicator of higher efficiency, probably due to improved cylinder filling and a more efficient combustion chamber that extracts more energy from a given fuel mass.

Guidelines for evaluating BSFC are well established and are frequently used to predict engine performance. One-half pound of fuel per horsepower per hour (0.50 BSFC) is the default norm for most calculations, but we can be more specific in many cases.

BSFC = observed HP ÷ observed fuel flow in pounds/hour

Or

Mass Fuel Flow = BSFC x anticipated HP

The following BSFC figures are typical of most modern performance engines:

0.48 to 0.55   Stock and medium performance engines
0.45 to 0.50   Performance engines with good heads
0.28 to 0.45   Most racing engines
0.35 to 0.38   Pro Stock style engines
0.55 to 0.65   Supercharged or turbocharged engines

Herein lies part of the problem with thinking of BSFC numbers as indicators of mixture ratios. At 0.37 BSFC a Pro Stock engine may be thought to be running too lean when, in fact, it is operating at the highest level of efficiency with a normal air/fuel ratio of about 13:1.

In contrast, supercharged engines run richer mixtures to complement boost pressure and discourage detonation. They absolutely run richer; not because they are inefficient, but to complement specific combustion characteristics inherent to boosted applications, not the least of which is charge cooling and the need for more fuel to augment the greater volume of air being supplied by the supercharging device.

When evaluating BSFC numbers, lower is almost always better (even when supercharged, where a 0.55 is still more efficient that a 0.65), as long as the combination supports safe combustion without detonation or overheating. It should be noted, however, that any engine still needs to run at the air/fuel ratio that produces best power. That's usually about 13:1 in normally aspirated en-gines and 11.6 to 12:1 in supercharged applications. You can't just run an engine lean and expect to get a low BSFC number. Tune for maximum torque and let the BSFC indicate how efficiently you generate that torque. At an indicated BSFC of 0.50 and engine burns 0.5 pounds of fuel per horsepower per hour (lb/hr). If the engine makes 500 hp, that's 250 lb/hr. If we're talking observed or uncorrected horsepower the fuel usage on your dyno sheet should indicate 250 lb/hr. So let's see how that works.

How to Read a Dyno Sheet, Part 2

One important thing to remember when you are looking at BSFC numbers on a dyno sheet is that BSFC relates only to observed (uncorrected) horsepower at the flywheel. Corrected horsepower is a calculated prediction based on specific correction standards for localized atmospheric conditions. Corrected horsepower figures multiplied by indicated BSFC will not match the recorded fuel flow in pounds per hour. This sometimes confuses novice tuners who are reading from the corrected figures on a dyno page. In many cases it leads them to think that the dyno is incorrect, or even worse, cheating them. I have seen numerous clients question power numbers that don't match BSFC numbers based on this incorrect assumption.

For component or system evaluation, any combination that increases BSFC is likely a poor choice and won't normally provide a power (torque) increase even though it doesn't necessarily effect a power loss. It simply means less efficiency and no gain with more fuel required to make the same power. When you hit upon a combination that lowers BSFC and increases VE, you're moving in the right direction. A running engine is, in effect, the ultimate flow bench. It telegraphs its preferences in terms of BSFC and VE. Appropriate power gains typically follow if the engine favors the current component mix.

Sharp dyno operators equipped with air flow turbines prominently display BSFC and VE numbers on screen so customers can observe them in real time during test sessions. When comparing components such as cylinder heads, it is often instructive to compare the graphs of indicated BSFC and VE to see how different components stack up against each other with all other factors remaining equal. VE improvements are frequently accompanied by BSFC improvements. If you build and tune for these two factors power will rise accordingly. Modern dyno software permits dyno data to be displayed and printed many different ways. A typical dyno sheet might show the following data streams:

**RPM • Corrected HP • Corrected TQ • BSFC
Fuel A • Fuel B • Oil Pressure • Water Temperature**

Many will also display an air/fuel ratio calculated from fuel and airflow measurements or, in some cases, a feed from an O_2 sensor. Most tuners make adjustments based on power gains or losses and the proximity of the air/fuel ratio to the ideal ratio for best power. Sounds reasonable, but for the purpose of further insight and evaluation, it might be more instructive to track the following critical data on a second page:

**RPM • TQ • A/F Ratio • BSFC • Airflow
Fuel Flow • VE • MAP • BMEP**

Note that all of the above closely model the engine's performance based on physical properties and internal combustion fundamentals. This is the data that reveals what is really going on inside the engine. In case you forgot, MAP is simply manifold absolute pressure and BMEP is the calculated brake mean effective pressure or an indicator of average cylinder pressure.

There are many different ways to look at dyno data. If you think about the processes involved in producing torque and horsepower, it becomes clear that a lot of valuable recorded dyno data is routinely overlooked. SuperFlow dynos and probably most other modern systems are able to offer clients a take-home version of the dyno software with all the recorded data for each test run. This allows you to use your own PC to study and evaluate all of your data and re-display it or graph it in ways that are meaningful to your goals.

If you are able to obtain your test data in this manner, I encourage you to take a close look at the information recorded and calculated on multiple data pages behind the primary page. Think about what all that data is telling you and how you might draw more informed conclusions by comparing it in graphic form. You'll be surprised at what you can learn about your combination.

A brief example might be a comparison between two cylinder heads with similar components, but different port flow characteristics. If one head shows better on the flow bench, you would expect to see a VE increase and a favorable BSFC if the combustion chamber provides supporting efficiency. Positive improvements in VE and BSFC translate directly to power gained at the flywheel. Closer examination of critical dyno data often reveals additional insight about where and how these gains occur in the power band and how they are affected by other contributing factors.

Fuel System Calculations

BSFC standards are also used to calculate anticipated fuel system requirements such as fuel pump capacity and fuel-injector size based on static flow rate. BSFC is a critical indicator of engine efficiency and it is one of your most valuable tools for evaluating engine performance. To gain further insight into the function of these systems and the calculations involved, we'll examine each of them separately.

Calculating Injector Size

Anticipated power levels are the basis for choosing an appropriate fuel injector size. The selection criteria are also based on injector static flow rate and the BSFC associated with the type of engine and its anticipated power level. Most injectors are static flow rated at 43.5 psi of fuel pressure or about 3 bar (1 bar represents 1 atmosphere, so 3 atmospheres would be 14.7 x 3 = 44.1 psi). Static flow is the flow rate achieved with the injector held wide open with steady-state fuel flow at the qualifying test pressure (43.5 psi). This provides a convenient method of flow rating different injectors, but it cannot be used in practice because the injector would remain on all of the time.

Injector on time is called the pulse width. It is the length of time in milliseconds that the injector is open and flowing fuel per intake event. The critical factor here is the cycle time available for individual fueling events per cylinder. The intake valve opens once every other revolution, and the cycle time for fueling diminishes with RPM. The cycle time is the amount of time it takes for two complete engine revolutions with one intake fueling opportunity during each cycle. Engine speed determines

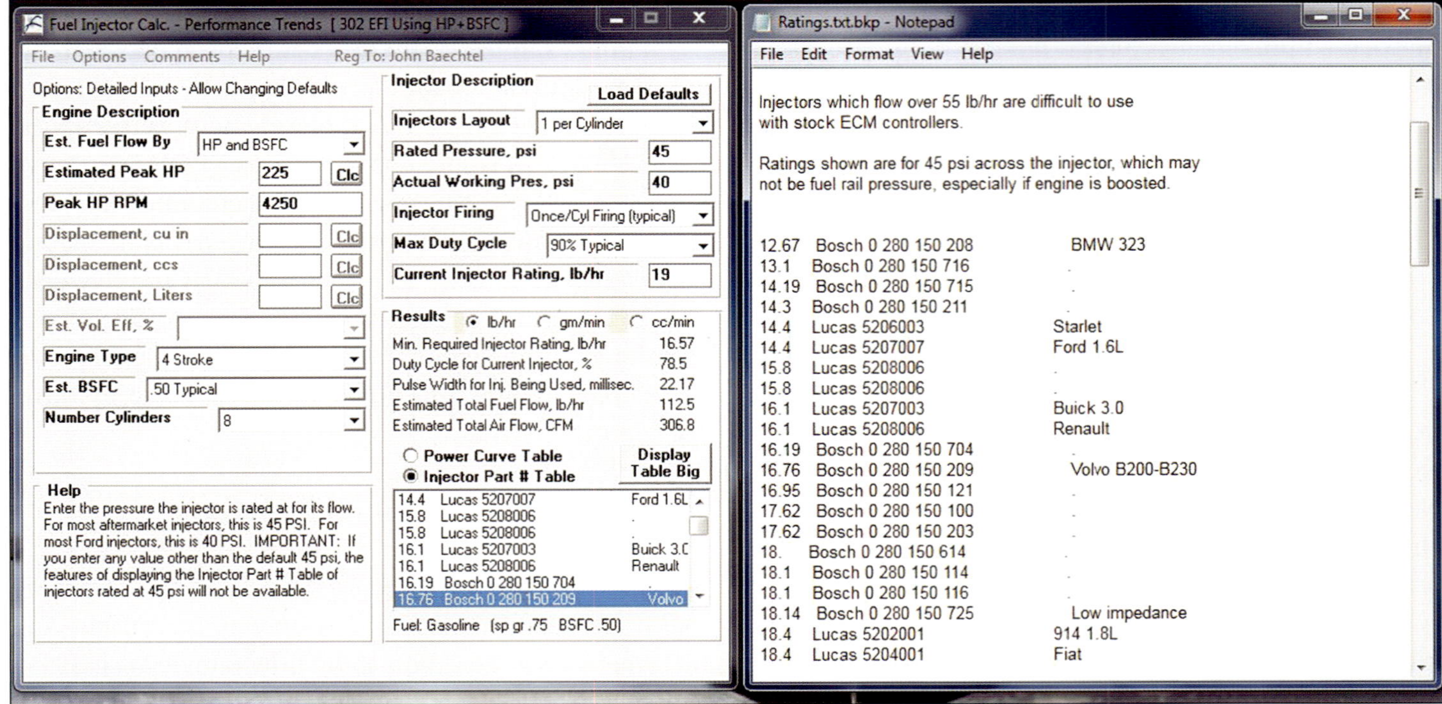

Performance Trends' Fuel Injector Calculator calculates injector size requirements, pulse width and duty cycle, pressure corrections, and all related fuel-injection calculations.

cycle time in milliseconds and is calculated with the following formula:

$$\text{Cycle Time} = 120{,}000 \div \text{rpm}$$

So, for a 6,000-rpm engine speed:

$$\text{Cycle Time} = 120{,}000 \div 6{,}000 = 20 \text{ milliseconds}$$

The ECU's calculated injector pulse width must fit within this narrow window of fueling opportunity. If the ECU were to calculate a pulse width of 20 milliseconds, that would equal 100-percent duty cycle, which the injector can't sustain because of injector head issues and the engine probably won't tolerate. Hence a safety factor is incorporated into injector size calculations. The safety factor is typically 80 percent of the injector's static flow rate, although some applications (Ford) use 85 percent. We'll call it 80 percent, which means the injector can be on up to 80 percent of the available time or duty cycle.

$$\text{Duty Cycle} = [(\text{injector pulse width} \times \text{RPM}) \div 120{,}000] \times 100$$

$$\text{Max Pulse Width} = 120{,}000 \div \text{RPM} \times \text{safe duty cycle (80\%)}$$

Example for 6,000 rpm:

$$\text{Max Pulse Width} = 120{,}000 \div 6{,}000 \times 0.8 = 16 \text{ ms}$$

$$\text{Duty Cycle} = [(16 \text{ ms} \times 6{,}000) \div 120{,}000] \times 100 = 80\%$$

If the commanded pulse width exceeds the safe duty cycle the injector static flow is inadequate for the application. If the difference is small, it can be compensated for with additional fuel pressure, but the best solution is usually to select an injector with a higher static flow rate.

This can get confusing because all injectors are not rated equally. While many aftermarket injectors are rated at 43.5 psi, Ford rates all their injectors at 39.15 psi, and most GM injectors are rated at 58 psi. The formulas all work the same, but you have to use the correct static flow rating for the injector you are considering. And its impedance must match the drivers in the ECU to prevent damage to the ECU because low-impedance injectors require more amperage to operate.

Factory systems such as Ford have used both mechanical and electronic returnless fuel systems to control evaporative emission since 1999. If you select injectors based on the factory rating and fuel pressure, you will satisfy all the requirements of the fuel system, which includes the ability of the Ford PCM to vary fuel pump voltage and pressure as required based on its knowledge of current operating conditions. When you have a choice of injectors rated at different flow rates, you can convert the flow rates at a different pressure delta with the following formula:

$$\text{Flow Rate at Desired Pressure} = \text{flow rate at old pressure} \times \sqrt{(\text{new pressure} \div \text{old pressure})}$$

Example: For a 30-lb/hr injector rated at 39.15 psi if you run it at 43.5 psi:

$$\text{New Flow Rate} = 30 \times \sqrt{(43.5 \div 39.15)} = 31.6 \text{ lb/hr}$$

Selecting an Injector

To size an injector we rely on the basic unit of efficiency, the brake specific fuel consumption (BSFC) at max power. For the purpose of calculation we use an estimated BSFC of 0.5 for normally aspirated engines and 0.65 for supercharged engines. We also incorporate the anticipated horsepower and a 20-percent safety margin, which gives an 80-percent operating duty cycle for the injector.

$$\text{Injector Flow Rate} = \text{BSFC} \times \text{HP} \div \text{number of injectors} \times \text{safe duty cycle of } 80\%$$

If we anticipate 500 hp and a BSFC of 0.5 we can calculate the flow rate that will safely support that HP level.

$$\text{Injector Flow Rate} = (0.5 \times 500) \div (8 \times 0.80) = 39.06 \text{ lb/hr}$$

A 39-pound injector will safely support 500 hp under all operating conditions. If, depending on your application, you need a specific type of injector (high or low impedance) and the only choices available are 36 or 42 pounds, select the larger injector to ensure the safety margin. If you are using the 39-lb/hr injector and dyno testing establishes that you are making 550 hp, you can preserve the safety margin by raising the fuel pressure accordingly. To calculate the required pressure increase we need to know the required flow rate to satisfy our safety margin at 550 hp.

$$\text{Injector Flow Rate} = (0.5 \times 550) \div (8 \times 0.8) = 42.96 \text{ lb/hr}$$

If the rated flow of the 39-pound injector was established at, say, 43.5 psi, we calculate the required pressure to satisfy the new horsepower requirement as follows:

$$\text{Fuel Pressure Change} = (\text{required flow rate} \times \text{rated pressure}) \div \text{current flow rate}$$

$$(42.96 \times 43.5) \div 39 = 47.9 \text{ psi}$$

If you raise the fuel pressure from 43.5 psi to 48 psi, the 39-lb/hour injector will support 550 hp at the safe duty cycle of 80 percent. You also need to enter the new flow rate into the ECU so it can make accurate calculations based on the new pressure. So if available injectors flow slightly less or more than required, small fuel pressure adjustments can bring them into line. The previous calculations can also be rewritten to determine the maximum safe horsepower that a given injector will support at its rated fuel pressure.

$$\text{Horsepower} = (\text{injector size} \times \text{number of injectors} \times \text{max duty cycle}) \div \text{BSFC}$$

To illustrate this examine the following examples taken from the Ford Racing catalog. Note that Ford uses a duty-cycle safety factor of 85 percent for its calculations.

$$\text{Naturally Aspirated at 0.5 BSFC} = (19 \text{ lb/hr} \times 8 \times 0.85) \div 0.5 = 258 \text{ hp at 85-percent duty cycle}$$

$$\text{Supercharged at 0.55 BSFC} = (19 \text{ lb/hr} \times 8 \times 0.85) \div 0.55 = 235 \text{ hp at 85-percent duty cycle}$$

$$\text{Supercharged at 0.65 BSFC} = (19 \text{ lb/hr} \times 8 \times 0.85) \div 0.65 = 199 \text{ hp at 85-percent duty cycle}$$

Injector Flow Rate (at 40 psi)	n/a HP (@ 0.5 BSFC)	Supercharged HP (@ 0.65 BSFC)
19 lb/hr	258	199
24 lb/hr	326	251
30 lb/hr	408	314
32 lb/hr	435	335
39 lb/hr	530	408
42 lb/hr	571	439
47 lb/hr	639	492
60 lb/hr	816	628

Note that these calculations are based on Ford injectors that are all rated at 39.15 psi. You can raise the pressure of any of these injectors to support more horsepower provided your fuel pump has the necessary capacity.

Fuel pressure is set with an adjustable fuel pressure regulator that has a vacuum line connected to the intake manifold. This allows the regulator to hold a constant pressure drop across the injectors as manifold vacuum varies according to throttle position.

The vacuum line from the manifold is connected to the reference side of the regulator diaphragm. It sees reduced pressure at idle or under cruise conditions. When the reference pressure drops, a corresponding drop in rail pressure occurs to keep the pressure drop across the injectors constant as manifold pressure fluctuates according to throttle and load.

The same thing occurs under boost pressure where each pound of boost causes a corresponding 1-pound increase in fuel rail pressure to maintain the pressure differential or "delta." If the fuel pump and fuel system are sized to maintain this 1:1 ratio according to manifold pressure, the flow rate of the injectors increase proportionately with demand to maintain the necessary fuel flow under full boost at WOT.

Fuel Pump Capacity and Flow Rate

As previously discussed, calculations with BSFC and horsepower reveal an engine's appetite for fuel in pounds per hour. Whatever that requirement might be, it represents the minimum steady-state fuel requirement at the injector or the carburetor. To ensure a positive reserve, fuel pumps are sized with plenty of capacity according to application.

To determine the required fuel pump flow rate for a given application, divide the calculated fuel requirement in pounds per hour by the nominal weight of gasoline, which is 6.009 pounds per gallon at a specific gravity of 0.72 and 65 degrees F. The specific gravity varies with temperature, but not enough to affect calculations for fuel pump capacity. There are two ways to calculate it: one for carburetors and one for injectors. The carburetor calculation typically indicates a lesser amount because it doesn't incorporate the safety margin applied to keep the injectors operating at their proper duty cycle.

$$\text{Flow Rate}_{inj.} = \text{injector size x number of injectors} \div \text{weight of gasoline (6.09 pounds per gallon)}$$

As noted in the previous example, 500 hp at a BSFC of 250 lb/hr. That represents the minimum requirement with the engine at WOT and maximum horsepower. To calculate the flow requirement, multiply the injector rating times the number of injectors and divide by 6.09 pounds.

$$\text{Flow rate}_{inj.} = (39 \times 8) \div 6.09 = 51.23 \text{ gallons/hour}$$

The calculation is slightly different for a carburetor:

$$\text{Flow rate}_{carb} = (\text{horsepower x BSFC}) \div 6.09$$

For 500 hp:

$$\text{Flow rate}_{carb} = (500 \times 0.5) \div 6.09 = 41.05 \text{ gallons/hour}$$

When you see fuel pumps rated above the indicated requirement, it's because the pump has to contend with the weight of the fuel in the fuel lines, g forces affecting the fuel mass, and fuel system restrictions such as injectors, carburetor needle and seats, fuel regulators, fuel filters, and the total number and type of fuel fittings in the entire fuel system. All of these factors conspire to restrict fuel capacity.

Accordingly, a good deal of overkill is required in pump selection. Bypass regulators are commonly used (except where previously noted) to reduce pumping effort while maintaining high-flow capacity. Pumps are required to supply the necessary volume at the appropriate fuel pressure, which can be anywhere from 20 to 70 psi for EFI systems and 5 to 12 psi for carburetors. A drag race car can generate substantial g force off the starting line; enough to literally stall the fuel column in the fuel line during peak acceleration. And supercharged engines create high fuel demands under boosted conditions with severe penalties if the fuel supply is inadequate. Accordingly, pump manufacturers build pumps with a substantial amount of reserve capacity to accommodate high fuel requirements. These calculations will help you determine the best pump.

If you know or have a good estimation of your car's performance capability, Barry Grant (BG) Fuel Systems recommends the following low-buck test to confirm adequate fuel delivery. This test applies to drag racing

Converting AN Sizes to Inches

AN stands for Army Navy and represents standard nomenclature for military-grade fittings. The AN size is the designated outside diameter (OD) of the metal tubing used with each size fitting. The assigned numbers relate to tubing size in 1/16-inch increments as shown below. Note that tubing and hoses have different wall thicknesses, so an AN designation doesn't necessarily reflect the exact ID of a particular tube or hose. AN thread sizes are SAE standard and are not compatible with inch-size pipe threads. To convert AN to inches, divide the fitting size by 16.

NPT stands for National Pipe Thread, and you'll find it on many AN-to-NPT adapters, where one end accepts a standard AN fitting and the other end is a common pipe thread used on fuel pumps, fuel filters, and other system components. The NPT chart indicates the closest AN fitting size for each common pipe thread based on the inside diameter (ID) or flow dimension. With the exception of adapters, AN threads are not compatible with NPT pipe threads. Note that the ID of a pipe fitting is reflected in the NPT designation.

Adapters usually list the AN size first as in -6AN, 3/8 NPT. Most popular tubing and hose manufacturers publish the ID of their tubing and hoses in their catalogs and online.

AN Fittings

AN Size	Tube Size OD	Thread Size (SAE)
-2	1/8	5/16-24
-3	3/16	3/8-24
-4	1/4	7/16-20
-5	5/16	1/2-20
-6	3/8	9/16-18
-8	1/2	3/4-16
-10	5/8	7/8-14
-12	3/4	1$\frac{1}{16}$-12
-16	1	1$\frac{5}{16}$-12
-20	1¼	1⅝-12
-24	1½	1⅞-12
-28	1¾	2¼-12
-32	2	2½-12

NPT Threads

Pipe Thread Size	Threads per Inch	Theoretical ID of Fitting	Closest AN Fitting Size
1/16	27	1/16	-3
1/8	27	1/8	-4
1/4	18	1/4	-6
3/8	18	3/8	-8
1/2	14	1/2	-10
3/4	14	3/4	-12
1	11½	1	-16
1¼	11½	1¼	-20
1½	11½	1½	-24
2	11½	2	-32

AN flair to pipe adapters allow AN lines to connect to components that only accept NPT fittings.

Carb adapters are another example of connectors that only accept NPT fittings.

Male and female AN couplers and unions are used to connect and extend AN lines.

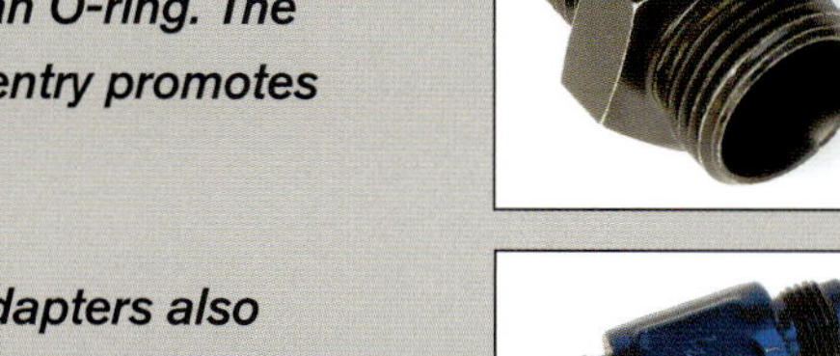

Fuel and oil systems often use radiused entry AN port adapters that seal with an O-ring. The radiused port entry promotes smoother flow.

AN and NPT adapters also include various pipe tee fittings and reducers to couple lines of differing sizes.

Fuel Flow Capacity Test

1/4-Mile ET (sec)	Time (sec) to Fill 1-Gal Can
7	Less than 12
8	15
9	20
10	25
11	30
12	35

cars, but it is well suited for other types as well since most other applications don't generate the high g-loading and instant fuel demand under acceleration.

BG suggests using an inexpensive 1-gallon gas can to check the capacity of your pump. Disconnect the fuel line or lines from your carburetor and route them from the regulator directly into the top of the open can. Switch on your pump and clock the time it takes to fill the can. Compare your results to the ones below. If your car is not a drag car, you should still strive for the quickest fill time possible, especially on cars that endure long high-gear pulls at max power, i.e., Bonneville or superspeedway cars. BG has pumps and fuel systems to satisfy all of these demands.

Note: Exercise caution. Perform this test in an open air environment away from open flames. Keep a fire extinguisher handy and don't try this alone.

One other thing to keep in mind for supercharged applications is the flow reduction that occurs when fuel pump pressure is increased under boost. This is also affected by available fuel pump voltage, which is frequently inadequate due to improper wiring or inadequate power source. This problem is most evident on EFI cars and must be addressed. Racing applications often datalog fuel pressure under maximum operating conditions to ensure that pump voltage and fuel rail pressure is adequate at all times. In most cases you have to generate this information yourself, but some manufacturers (Ford) have fuel pump maps that show flow rate versus delivery pressure at a given voltage.

Fuel System Formulas at a Glance

BSFC = observed HP ÷ observed fuel flow in pounds/hour Or Mass Fuel Flow = BSFC x anticipated HP

Cycle Time = 120,000 ÷ RPM

Duty Cycle = [(injector pulse width x RPM) ÷ 120,000] x 100

Max Pulse Width = 120,000 ÷ RPM x safe duty cycle (80%)

Flow Rate at Desired Pressure = flow rate at old pressure x $\sqrt{\text{(new pressure} \div \text{old pressure)}}$

Injector Flow Rate = BSFC x HP ÷ number of injectors x safe duty cycle of 80%

Fuel Pressure Change = (required flow rate x rated pressure) ÷ current flow rate

Horsepower = (injector size x number of injectors x max duty cycle) ÷ BSFC

$\text{Flow Rate}_{inj.}$ = injector size x number of injectors ÷ weight of gasoline (6.09 pounds per gallon)

Flow rate_{carb} = (horsepower x BSFC) ÷ 6.09

ATMOSPHERICS AND COMBUSTION MATH

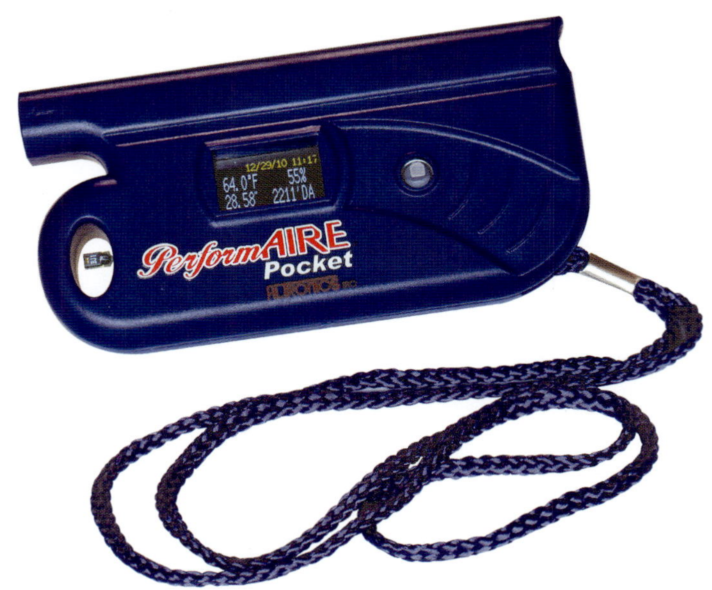

Affordable trackside weather stations like Altronic's PerformAIRE Pocket unit are common at all levels of racing from Pro Stock to Saturday night bracket racing.

In addition to the nuts-and-bolts math of engine design and assembly, it is useful to know a few things about combustion and the effects of atmospheric pressure on engine tuning and performance. In particular, I discuss air/fuel ratios, correction factors for dyno testing, and the effects of altitude on engine performance. There are only a few formulas in this chapter, but they are important to your understanding of how your engine operates and how it should be tuned.

Atmospheric Effects on Engine Performance

Engines deliver their best performance at sea level because that's where atmospheric pressure is greatest and air den-

sity is highest, notwithstanding atmospheric variations. The common reference for atmospheric conditions is standard temperature and pressure (STP) or the "standard day." STP is defined as 60 degrees F, 29.92 barometric pressure, and dry air (zero humidity). As the components of STP vary according to weather, air density also changes for better or for worse. By these standards, STP equals 100-percent air density, but zero density altitude. As temperature, barometer (atmospheric pressure), and humidity vary, air density fluctuates accordingly.

Each 5-degree change in temperature (from STP) equals a 1-percent change in air density. Hence a 90-degree day would have an air density of 94 percent if other factors aren't considered. The change is not entirely linear since it

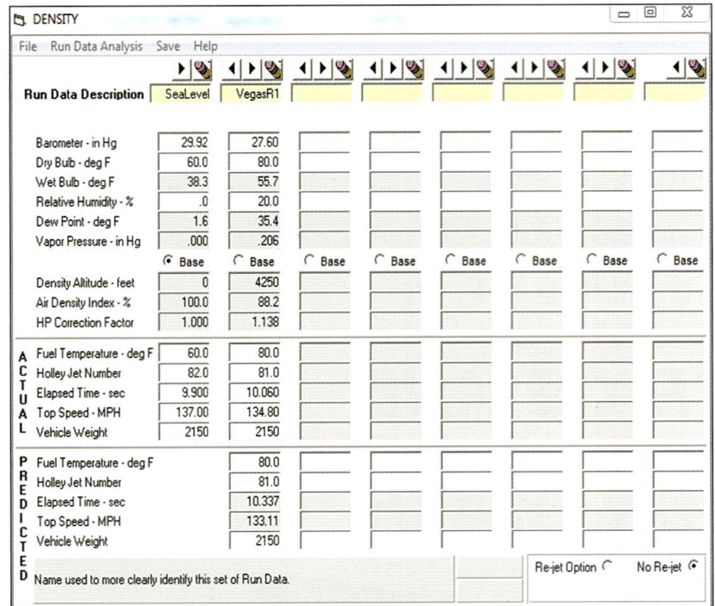

Racing Systems Analysis' DENSITY program is powerful software that performs complex calculations for accurate weather corrections at the track. It looks at your particular motorsports application and makes calculations based on elevation, barometer, temperature, humidity, dew point, vapor pressure, fuel type, carbs or injectors, naturally aspirated or supercharged, jet size, and mechanical injector data. It also provides a run data analysis with appropriate corrections.

is always accompanied by barometer and humidity changes. Still, it's close. Temperature typically has a greater effect than humidity; something on the order of nearly 2 to 1. The effect of barometer is linear with density following by percentage. If the barometer drops by 1 percent, air density falls by 1 percent. Humidity is more complicated because of the relationship between temperature and water vapor. Air is capable of holding more water as temperature rises, hence 50-percent humidity at 90 degrees F causes almost double the loss of air density as the same percentage of humidity at 60 degrees F or STP.

These effects are interrelated. If you're running at a higher altitude and the barometer falls 2 percent to, say, 29.32, you lose 2-percent air density. Then the temperature shoots up to 90 degrees F and you lose another 6 percent. And at 50-percent humidity you lose another 2 percent. Now you've encountered a 10-percent drop in air density and, based on the widely accepted power to density ratio of 75 to 80 percent, you're looking at about an 8-percent loss

of power. For your convenience, the NHRA's published correction factors for elapsed time and speed based on elevation change are listed on page 116. Multiply your ET and speed by the appropriate conversion for your elevation to determine your projected performance at sea level.

Density Altitude

Density altitude can be thought of as altitude measured in terms of air density rather than distance. It is basically the pressure altitude adjusted for nonstandard temperature and humidity. When temperature and humidity increase, the density altitude of a given location may be substantially higher than the actual elevation in feet. For example, at Bonneville, you might be racing at 4,200 feet elevation, but the density altitude might be 6,500 feet according to atmospheric conditions. Hence you are racing at 6,500-foot conditions and appropriate power loss can be expected. Here is the standard formula for estimating power loss with altitude:

$$HP_{loss} = (\text{elevation} \times 0.03 \times \text{HP at sea level}) \div 1{,}000$$

For 800 hp running at 3,500 feet:

$$(3{,}500 \times 0.03 \times 800) \div 1{,}000 = 84\text{-hp loss}$$

This formula yields an estimate that does not account for the displacement of fuel molecules by water vapor in the intake air supply. For greater tuning accuracy, you may want to get a performance weather station to help you account for all contributing factors. It makes jetting and air bleed changes more accurate based on ambient conditions.

Optimum conditions for engine performance include high pressure (barometric or ram), low temperature, and minimal water-vapor pressure. When these combine in the right proportions, air density increases and more tightly packed oxygen molecules are available to combine with fuel for combustion. From a tuning perspective, note that excessive vapor in the air also tends to cool the burn and further reduce power. For any water vapor pressure below 40 to 45, percent the engine often responds to slightly richer jetting. Higher percentages displace more oxygen molecules and may require less fuel and more timing to burn the cooler, wetter mixture.

Performance Trends' Weather Wiz program uses your input data to make precise weather and tuning calculations that affect engine, vehicle, and carburetor performance.

Performance Weather Stations

In recent years handheld units have grown in popularity (the PerformAire unit is a favorite). They help you make tuning decisions based on accurate measurements of station pressure (barometric pressure at the track), air temperature, and the water vapor content in the air.

Performance weather stations vary in content and to some degree methodology, but they all strive to accomplish the same thing, which is to provide the racer with accurate atmospheric data so he can make appropriate tuning adjustments for race day conditions (see sidebar "Optimum Fuel Mixture Ratios"). Racers have traditionally used air-density gauges to monitor density changes as the air changes during a given day, but air-density gauges do not account for humidity or vapor pressure. They simply indicate the percentage based on ambient temperature and pressure.

If you have ever taken a flight out of the Denver airport, you may have noticed nervously that the takeoff roll seems about 50 percent longer and farther than what you are used to at most airports. There is less air density to support the plane's wings; hence it requires a long roll and higher speed before the pilot can lift off. In the early days of aviation, piston-engine-aircraft performance diminished as altitude increased. Once supercharging was employed to artificially raise the air density, aircraft performance increased proportionately. The same lack of density takes away performance from your engine on a percentage basis. Any decrease in pressure or increase in temperature or vapor pressure raises the density altitude and reduces performance. (See Appendix B "Handy Conversion Factors.")

Shame on the Weatherman

Your local weatherman is not a reputable source for racing weather data. He may quote a fabulous day with a pressure reading well above 30, but your uncorrected mercury barometer tells you different—say, 26.10, for example. Why? Well, it's because the National Weather Service

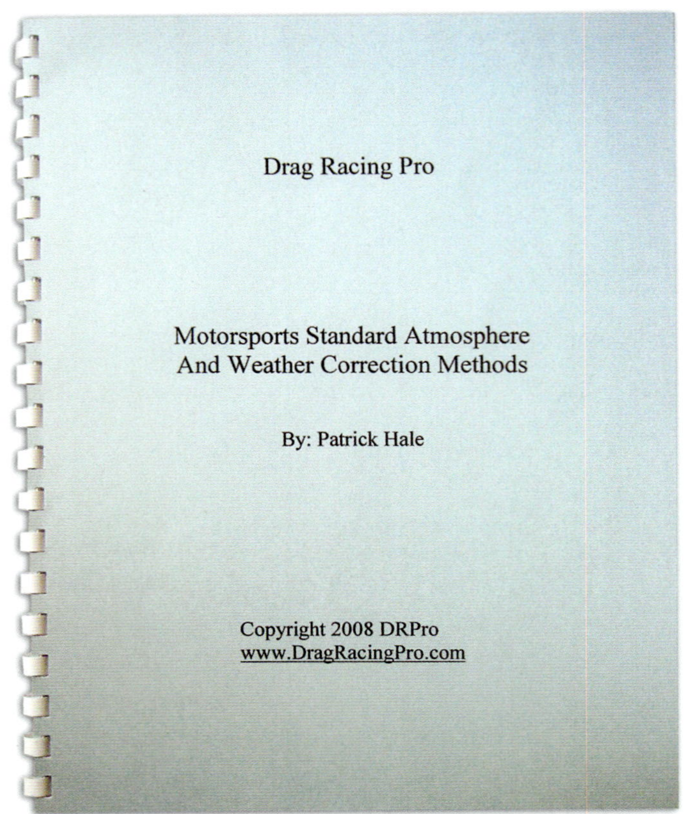

Patrick Hale's **Motorsports Standard Atmosphere And Weather Corrections Methods,** *available through his Web site www.DragRacingPro.com, is pretty much the last word on the subject of racing weather corrections. Hardcore racers using it along with a PC weather program or a handheld unit achieve great results.*

corrects all altitude based conditions to known sea-level conditions for equivalent temperature and pressure. Hence it is critical that you use uncorrected (local) station pressure for your calculations or input to your weather calculator. In a pinch you can get pretty close if there is a nearby airport. Call and ask the tower for an uncorrected station pressure reading. This will be the actual atmospheric pressure at your location without corrections for temperature and water vapor.

It's also important to use your own weather station equipment when visiting the dyno. Ask the operator to give you the uncorrected barometric pressure that he enters into the dyno software along with the vapor pressure and the specific gravity of the fuel. Make certain that your printouts or the data disc the operator gives you after the test include all observed (uncorrected) figures, which are what the engine actually delivered on that given day. And be certain that you have numbers for the inlet air temperature. There are few unscrupulous dyno operators, but if your operator doesn't spend at least 20 minutes taking weather measurements and entering them into the dyno software, it would be a good idea to ask why. Observed numbers should match your BSFC numbers and still give you an accurate picture of engine performance even though they don't seem as impressive as corrected numbers. That's why there is no point in comparing your numbers to someone else's engine run under different conditions.

The Motorsports Standard Atmosphere

Motorsports engineering authority Patrick Hale has compiled the results of more than 35 years of racing and motorsports success into *Motorsports Standard Atmosphere And Weather Correction Methods*, a book available through his Web site www.DragRacingPro.com. If you are serious about motorsports tuning and weather corrections, it is pretty much the last word on the subject. In it Hale defines the motorsports standard atmosphere in relation to aerospace and general automotive atmospheric definitions and discusses the intricacies of atmospheric temperature and pressure, measurement techniques, density altitude, water vapor displacement, SAE corrections, corrections for methanol and nitromethane fuels, drag racing ET and MPH corrections, aerodynamics, wind corrections, and more. It's heady stuff and beyond the scope of this book, but not so deep that you can't grasp it with a little effort. For those determined to tackle advance tuning issues and the math behind them, I highly recommend it.

Lambda and Air/Fuel Ratios

Lambda and air/fuel ratios are often confused, but the relationship is really quite simple. Air/fuel ratio indicates the mixture ratio of air and fuel. A 12:1 air/fuel ratio indicates 12 parts (or pounds) air to 1 part (or 1 pound) fuel. Because it is a ratio, it can be anything you want. It could be 12 pounds of air to 1 pound of fuel, for example. Calibration engineers prefer Lambda because it gives them an easy-to-read percentage of the air/fuel mixture's deviation from the ideal ratio for complete combustion.

The ideal ratio is called stoichiometric, and it is defined as an air/fuel ratio of 14.68:1 (for gasoline), or Lambda equals 1.00. They both represent the same mixture ratio. The stoichiometric ratio is the control ratio for all modern electronically managed engines and the tuning standard for EFI calibration. This represents the chemically "correct" ratio where emission components are at their lowest and should not be confused with a best power ratio that is closer to 13:1.If lambda represents the stoichiometric mixture, anything greater than 1 represents a leaner mixture and anything less than 1 indicates a richer mixture. In the performance world we know that best power (torque) usually occurs at an air/fuel ratio of about 13.2:1 ± 0.2. The equivalent Lambda is 0.9.

$$\text{Lambda } (\lambda) = \text{indicated air/fuel ratio} \div \text{stoichiometric air/fuel ratio}$$

$$13.2 \div 14.68 = 0.899 \text{ or } 0.9 \text{ Lambda}$$

Lambda is defined as an excess air ratio, but it really works both ways. In the preceding example, 0.9 Lambda indicates the engine is using only 90 percent of the air required to achieve stoichiometric combustion. It is running rich to achieve best power (torque). The slightly richer mixture optimizes flame travel and drives the mixture toward a faster reaction rate. When combined with the appropriate quench to produce good mixture turbulence, combustion efficiency and power rise accordingly.

With all modern performance engines moving to electronic management and fuel injection, enthusiasts would do well to start thinking in terms of Lambda instead of air/fuel ratio. If you can remember that 13.2:1 is usually best for peak power, it's just as easy to remember that 0.9 Lambda is the same thing. If you're going to learn to calibrate EFI systems, it's good to get comfortable with Lambda. (For an in-depth look at Lambda in the tuning process, see Greg Banish's excellent SA-Design books, *Designing and Tuning High-Performance Fuel Injection Systems* and *Engine Management: Advanced Tuning*, published by CarTech.)

Dynamometer Correction Factors

Common testing procedures can be confusing if you don't understand where the numbers come from and how the local testing environment affects them. It is important to recognize that a dyno is basically a sophisticated data acquisition system that measures degrees of change (good or bad) caused by various components, modifications, or tuning changes with accommodation to ambient conditions. A dyno provides raw data that is typically corrected to accepted standards for comparison. If you're testing cylinder heads for example, you don't necessarily need corrected numbers if your tests are all performed under the same conditions. Observed torque either rises or falls from your baseline, and you are likely to learn more about the head by comparing VE and BMEP. Still, it is important that you understand the correction process and how it can affect your data.

Observed torque is the raw torque value measured by the load cell (strain gauge) on the dyno. It's basically "what you see is what you get." Of course the engine's performance is affected by local atmospheric conditions and that is what you see. Correction factors (CF) are used as benchmarks for comparison purposes. If you're testing on the same day and conditions remain stable, you can judge performance by observed figures and evaluate efficiency changes by comparing relevant data such as VE. If you're testing on a different day with new conditions, a correction standard becomes useful. Accordingly it is helpful to know the testing landscape and the terms that define it.

Observed horsepower is calculated from observed torque using the formula discussed in Chapter 5:

$$HP = (\text{observed torque x RPM}) \div 5{,}252$$

Corrected torque and horsepower are the observed figures multiplied by the correction factor. Correction factors are based on observed inlet air temperature, wet bulb temperature, and barometric pressure. Wet bulb temperature is used to calculate the inlet air's vapor pressure and humidity. A corrected barometric pressure is calculated by subtracting the corrected vapor pressure. Cool air and lower vapor pressure allow for denser (oxygen rich) air and less fuel displacement by water vapor. The dyno measures most of this on the fly and makes the calculations for you. There are two basic correction factors, as described next.

Correction Factor SAE J607

This factor is commonly used by the performance industry, particularly on engine dynos. It corrects observed data to STP, or 60 degrees F at a barometric

NHRA Altitude Correction Factors

Elevation (feet)	ET Factor	MPH Factor
1,200	0.9874	1.0129
1,300	0.9861	1.0143
1,400	0.9848	1.0157
1,500	0.9835	1.0171
1,600	0.9822	1.0185
1,700	0.9809	1.0199
1,800	0.9796	1.0213
1,900	0.9783	1.0227
2,000	0.9770	1.0241
2,100	0.9757	1.0255
2,200	0.9744	1.0269
2,300	0.9731	1.0283
2,400	0.9718	1.0297
2,500	0.9705	1.0311
2,600	0.9692	1.0325
2,700	0.9679	1.0339
2,800	0.9666	1.0353
2,900	0.9653	1.0367
3,000	0.9640	1.0381
3,100	0.9627	1.0395
3,200	0.9614	1.0409
3,300	0.9601	1.0423
3,400	0.9588	1.0437
3,500	0.9575	1.0451
3,600	0.9562	1.0465
3,700	0.9549	1.0479
3,800	0.9536	1.0493
3,900	0.9523	1.0507
4,000	0.9510	1.0521
4,100	0.9497	1.0535
4,200	0.9484	1.0549
4,300	0.9471	1.0563
4,400	0.9458	1.0577
4,500	0.9445	1.0591
4,600	0.9432	1.0605
4,700	0.9419	1.0619
4,800	0.9406	1.0633
4,900	0.9393	1.0647
5,000	0.9380	1.0661
5,100	0.9367	1.0675
5,200	0.9354	1.0689
5,300	0.9341	1.0703
5,400	0.9328	1.0717
5,500	0.9315	1.0731

pressure of 29.92 in Hg and dry air (zero humidity). It subtracts corrected vapor pressure from observed barometric pressure to correct for water vapor in the air.

$$CF = (29.92 - \text{corrected barometric pressure})^{1.2} \times [(\text{observed inlet temp} + 460) \div (520)^{0.6}]$$

The resulting correction factor is multiplied by observed torque and horsepower to obtain corrected figures. Note that temperature in this formula is converted to degrees Rankine, and that this standard yields power numbers approximately 4 percent higher than SAE1349.

Correction Factor SAE 1349

This correction factor is standard to the OEM auto industry and is frequently the standard for chassis dyno work, although most engine and chassis dyno software automatically calculates both. This standard converts to 77 degrees air temperature, 29.31 barometric pressure, and includes a factor for 85-percent mechanical efficiency.

$$CF = 1.180 \times [(990 \div Pd) \times (Tc + 273 \div 298)^{0.5}] - 0.18$$

Where:
CF = final correction factor multiplier
Pd = pressure of dry air in hPa (990 hPa = 99kPa)
Tc = air temperature in degrees Celsius

Since most dyno testing is comparative, either standard can serve your needs. For that matter even observed numbers are instructive as long as you maintain consistent comparisons. The important thing is to choose a standard and stick with it for all your testing requirements.

NHRA Altitude Correction Factors

Multiply your ET or speed by the indicated correction factor in the adjacent chart to estimate your performance at sea level. These corrections are based on standard conditions and can be used as a general rule. Actual vehicle performance may vary according to specific local weather conditions.

Fuel Specifications

Fuel quality and fuel mixture ratios are an important component of engine performance. In addition to juggling air/fuel ratios for best performance and optimum fuel economy, tuners are also concerned with fuel heat values, burn rates, octane, and the density or specific gravity of various fuels. For example, fuels with a higher specific gravity affect flow through the jet and alter tuning. The accompanying fuel/mixture ratio and fuel quality specifications are provided for reference. Note in particular the octane differences and the heat values spread between regular and premium gasoline and E85 Ethanol. Also note the difference in stoichiometric ratios between fuels.

Optimum Fuel Mixture Ratios

Mixture Ratio	A/F Ratio	Lambda
Stoichiometric	14.7	1.00
Max Power	12.6 –13.2	0.86 – 0.89
Max Power (Supercharged)	11.0 –11.6	0.75 – 0.79
Best Economy	15.0 –15.4	1.02 – 1.05

Lambda values are easily determined by merely dividing the air/fuel ratio by the stoichiometric (14.7). As an example, a rich mixture of 12.5:1 divided by 14.7:1 = 0.85 Lambda. Another example would be a lean mixture of 15.0:1 air/fuel ratio which converts to 15:1/14.7:1 = 1.02 Lambda.

Fuel Quality Specifications

Fuel	Lower Heating Value kJ/kg	Octane (R + M) / 2	Density kg/l	Stoichiometric Ratio kg/kg
Gasoline Regular	42.7	87	0.740	14.8
Gasoline Premium	43.5	93	0.755	14.7
E85	29.19	101.6	0.783	9.87
Ethanol	26.8	104.2	0.790	9.00
Methanol	19.7	104.5	0.790	6.47
Propane	46.3	104.5	0.510	15.67
Diesel	42.5	25	0.835	14.5
LPG	43.1	110	0.540	15.5

This chart indicates the heating value, octane rating, density, and effective stoichiometric ratio for commonly available fuels.

Atmospheric/Combustion Formulas at a Glance

$$HP_{loss} = (\text{elevation} \times 0.03 \times HP \text{ at sea level}) \div 1{,}000$$

$$\text{Lambda} = \text{current air fuel ratio} \div 14.68$$

Dyno Conversion Factors

$$\text{SAE J607 CF} = (29.92 - \text{corrected barometric pressure})^{1.2}$$
$$\times [(\text{observed inlet temp} + 460) \div (520)^{0.6}]$$

$$\text{SAE 1349 CF} = 1.180 \times [(990 \div Pd) \times (Tc + 273 \div 298)^{0.5}] - 0.18$$

CAMSHAFT MATH

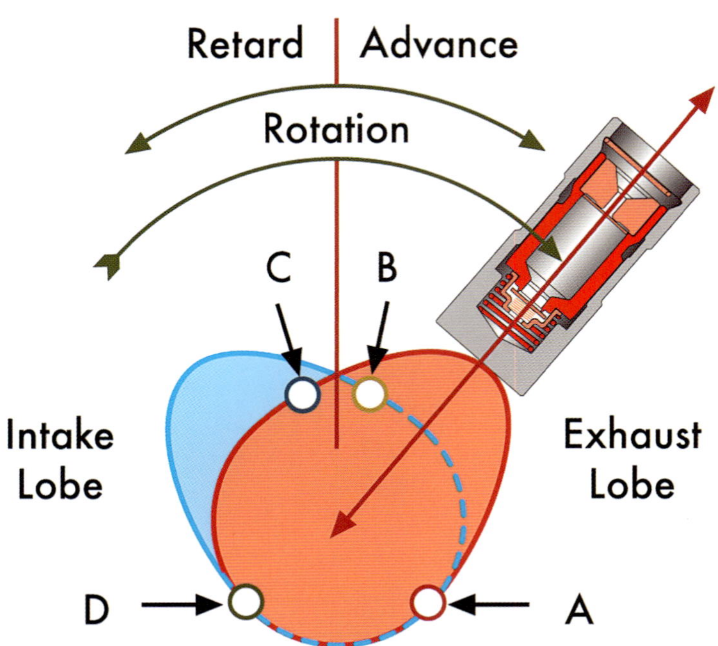

A. Exhaust Valve Opens - Power Stroke
B. Intake Valve Opens - Exhaust Stroke
C. Exhaust Valve Closes - Intake Stroke
D. Intake Valve Closes - Compression Stroke

The basic cam lobe shape is an eccentric with the lifter riding on the base circle. As the cam rotates, the lifter moves up the flank of the lobe and over the top to open the valve. (Courtesy Comp Cams)

Camshafts are a subject that all car guys love to talk about, and those who listen tend to roll their eyes at the staggering number of terms and numbers. There's lift and duration, overlap and lobe centers, lobe separation angles and so on, and all of them have numbers attached. Most enthusiasts have a basic understanding of the cam, lifter, pushrod, rocker arm, and valve relationship, but keeping all the numbers straight is often intimidating. This chapter looks at the basic calculations relating to camshafts and valvetrain components and how you can use them to equip and tune your particular combination for top performance.

Lobe Centers

Imagine a line passing from the center of the cam directly through the highest point on the cam lobe. This is the geometric centerline of that particular lobe. To avoid confusion when comparing cams, remember that the lobe center angle is measured between the centerlines of the intake and corresponding exhaust lobes, while lobe centerline is the angle measured between the center of the lobe and TDC. Lobe center angle is fixed and cannot be changed after the cam is ground. The lobe centerline can be altered by advancing or retarding the cam. When

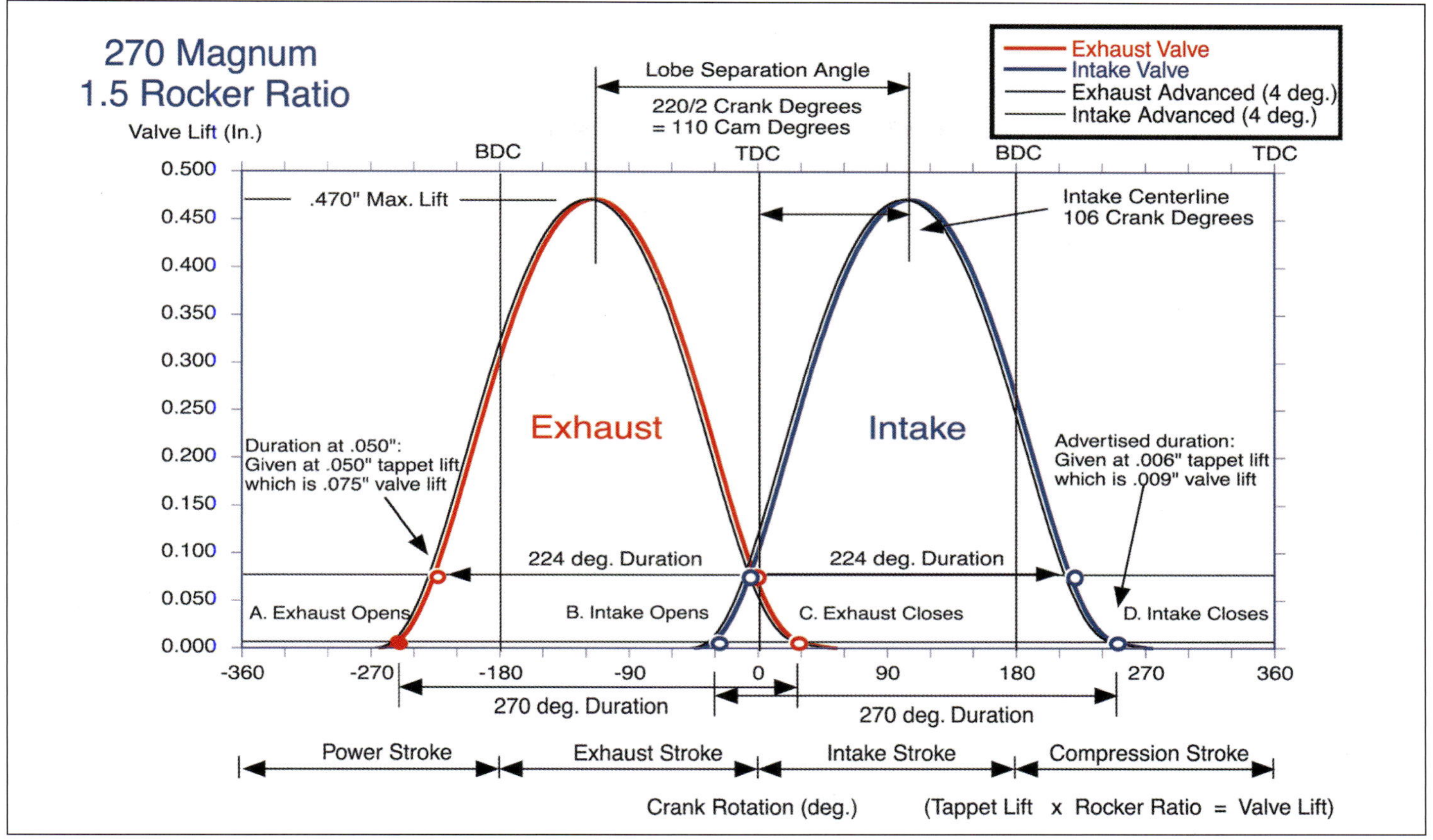

This illustration tells you everything you need to know about camshaft operation. In addition to lift duration and overlap, it also indicates the intake and exhaust centerlines, lobe separation angle, and the location of timing events relative to the four cycles. Note that 0.050-inch tappet lift equals 0.050 times the rocker ratio to indicate valve lift at the checking point. (Courtesy Comp Cams)

you do this you are effectively moving the intake lobe centerline closer to or farther from TDC.

In terms of engine performance, the lobe center angle is significant. A larger angle yields less valve overlap (the period when both valves are open at the same time). This permits the cylinder to begin building pressure sooner and that boosts low-speed torque. Decreasing the angle crates greater overlap and moves the torque curve higher in the RPM range, effectively narrowing the engine's powerband.

For most street applications always select a cam that builds as much torque as possible. Generally you want valve events that produce a wider lobe center angle, decreasing valve overlap. One of the advantages of the new high-velocity street roller profiles is that you maintain good idle quality by using wider 112- to 115-degree lobe center-lines, but you also have a high-RPM boost with more aggressive lobe profiles (increasing effective duration). The result is a broad torque curve ideal for street use.

Supercharged or turbocharged applications should avoid cams with excessive overlap because the pressurized intake system already provides effective cylinder filling and forced exhaust scavenging. In these applications, long overlap can be detrimental because some of the intake charge can be blown right through the engine without being burned. For the average street and strip enthusiast, all of these factors are taken care of by the cam manufacturer. Their vast experience lets them provide you with a cam that they know will work for you application.

Understanding Cam Specs

For the purpose of this discussion I will speak in terms of opening and closing valve events. Intake opening (IO) and exhaust opening (EO) represent the intake and exhaust opening points in crankshaft degrees. Intake

```
COMP CAMS
ENGINE SMALL BLOCK CHEVY 265-400
PART # 12-521-5        08/13/2003
GRIND NUMBER CS 41/15 H6

                   INTAKE    EXHAUST
VALVE ADJUSTMENT   HYD HYD
GROSS VALVE LIFT    .420        .420
DURATION AT
 .006  TAPPET LIFT  297        299
VALVE TIMING      OPEN      CLOSE
AT .050 INT        17 BTDC     49 ABDC
        EXH        51 BBDC   19 ATDC
THESE SPECS ARE FOR CAM INSTALLED
AT 106   INTAKE CENTER LINE
                   INTAKE    EXHAUST
DURATION AT .050    246        250
LOBE LIFT          .2800      .2800
LOBE SEPARATION    106.0

981-16        SPRINGS REQUIRED.
VALVE SPRING SPECS FURNISHED WITH SPRINGS
```

The camshaft timing card included with your cam provides the essential specifications for installing your cam correctly. It includes lobe lift, net valve lift, timing points at the advertised duration, duration at 0.050-inch lift, lobe separation angle, and the installed intake centerline. Some cards also include overlap and valvespring specs. Most manufacturers now publish all their standard cam cards online for your convenience.

closing (IC) and exhaust closing (EC) are the intake and exhaust closing events. Cam cards publish these points based on the manufacturer's chosen reference points: typically 0.006 inch for advertised duration and 0.050 inch for a universal checking reference based on an agreed amount of lobe lift where reasonable flow is initiated. The following formula is used to calculate intake and exhaust duration. It applies to any lift as long as your cam card specifies opening and closing figures for a particular lift value.

$$\text{Duration at Specified Lift} = \text{opening point} + 180 \text{ degrees} + \text{closing point}$$

For example, a COMP Cams XE274H-10 hydraulic cam lists the following opening and closing points for a checking lift of 0.006 inch:

IO = 31-degrees BTDC   IC = 63-degrees ABDC
EO = 77-degrees BBDC   EC = 29-degrees ATDC

Hence,
Intake Duration = 31 + 180 + 63 =
    274 degrees at 0.006-inch lift
Exhaust Duration = 77 + 180 + 29 =
    286 degrees at 0.006-inch lift

From this you can calculate the intake and exhaust centerlines. To find the intake lobe centerline, divide the intake duration by two and subtract the indicated intake opening point as shown in our example with the Comp XE274 example.

$$\text{Intake Centerline} = (\text{duration} \div 2) - \text{IO}$$

$$\text{Intake Centerline} = (274 \div 2) - 31 = 106 \text{ degrees}$$

Sometimes you find a very mild or stock cam where the IO occurs after TDC (ATDC). In this case just add the IO figure to one half of the duration.

On the exhaust side the formula is similar, but instead of subtracting the intake opening point, subtract the exhaust closing point.

$$\text{Exhaust Centerline} = (\text{calculated duration} \div 2) - \text{EC}$$

$$\text{Exhaust Centerline} = (286 \div 2) - 29 = 114 \text{ degrees}$$

Once you know this it's easy to calculate the lobe separation angle (LSA) which is the difference between the two centerlines. Simply add the calculated centerlines together and divide by 2.

$$\text{Lobe Separation Angle} = (\text{intake centerline} + \text{exhaust centerline}) \div 2$$

$$\text{LSA} = (106 \text{ degrees} + 114 \text{ degrees}) \div 2 = 110 \text{ degrees}$$

If a cam is ground "straight up," both centerlines are the same and the LSA is one half of their sum. More commonly you find that cam companies grind their street cams 4 degrees advanced to help boost low-speed torque on longer-duration cams. You can see this in the Comp XE274 example where the intake centerline is 106 degrees, but the LSA is 110 or 4 degrees advanced. Note that 110 degrees is exactly halfway between 106 and 114 degrees. This practice moves the IC event 4 degrees ahead, which tends to diminish top end power in favor of more low-speed grunt for street engines.

One other point to note is the use of parenthesis around some timing points. This notation indicates that the cam actually closes the valve after TDC instead of before, even though the card indicates BTDC. You only find this on short-duration cams, but it is important to note if you're making calculations with a small cam.

Calculating Valve Lift

Net valve lift is a function of camshaft lobe lift and rocker arm ratio. Lobe lift (sometimes called cam rise) is the height of the eccentric portion of the cam lobe above the base circle. The rocker arm transfers the motion of the valve lifter riding on the cam lobe to the valve and increases the lobe lift by the amount of the rocker ratio, which is typically 1.5 to 1.7:1. It provides a convenient means of increasing valve lift without a space or packaging penalty. This is very evident in a pushrod engine where the valvetrain is compact and easily packaged compared to the complication and excessive size required for single and double overhead cam arrangements.

Net valve lift differs according to the type of lifter. To accommodate thermal expansion, clearance is build into the system in the form of clearance ramps and valve lash for mechanical (solid) lifter cams. The valve lash clearance must be subtracted from the total valve lift to obtain the net valve lift for this type of cam.

Mechanical Lifter Cam Net Lift =
(lobe lift x rocker ratio) − valve lash

Example: For a Lobe Lift of 0.300 inch and a 1.5:1 rocker ratio with a 0.022-inch valve lash:

Net Lift = (0.300 x 1.5) − 0.022 = 0.428 inch

A hydraulic camshaft automatically adjusts for thermal expansion via lifter preload against an internal hydraulic plunger. No clearance is necessary and these lifters are typically adjusted with a specified amount of preload or a preferred degree of turn from zero lash; usually one-quarter to one-half turn down. In this case

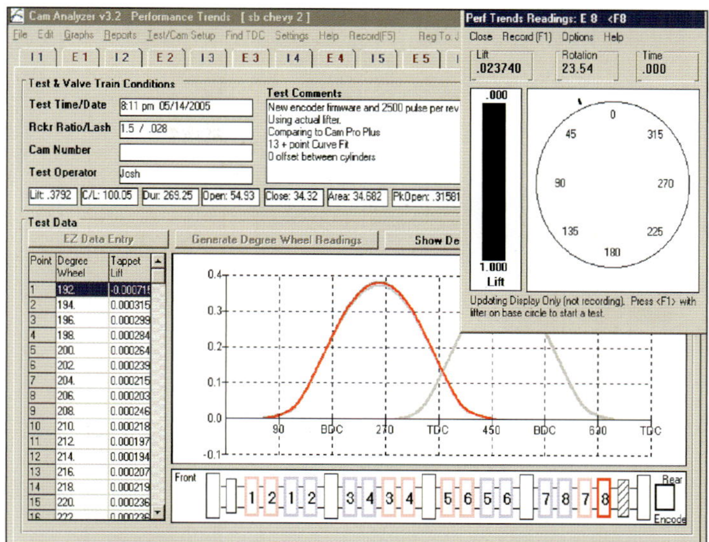

The Cam Analyzer from Performance Trends uses hand-measured cam profiles or computerized file formats such as Cam Doctor and Cam Pro Plus to provide detailed camshaft lobe evaluation for use in performance simulations.

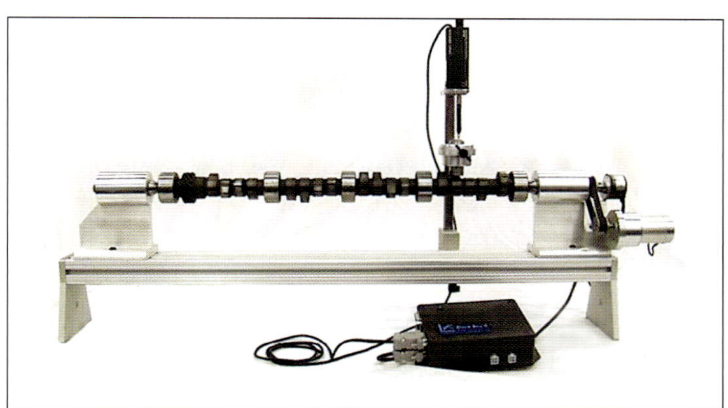

Cam Analyzer uses its motorized unit to rotate the cam so the profiler can read the lobe.

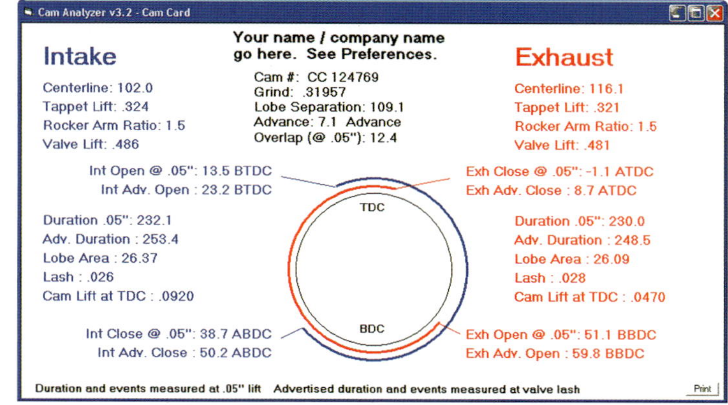

Cam Analyzer generates its own cam card for your reference files.

Degreeing the cam requires accurately locating TDC with a degree wheel and pointer, and a dial indicator set up to read lifter travel as you rotate the engine.

the net valve lift is based on the lobe lift and the rocker ratio alone.

$$\text{Net Lift} = \text{lobe lift} \times \text{rocker ratio}$$

$$\text{Net Lift} = 0.300 \times 1.5 = 0.450 \text{ inch}$$

Mechanical (solid) cams are typically smaller than their hydraulic counterparts due to loss of lift attributable to valve lash. But mechanical cams, unlike hydraulic cams, can be tuned somewhat by altering valve lash. Tightening the lash adds lift and starts the valve event sooner, effectively mimicking a larger cam. To accommodate various tuning changes, this is often limited to either the intake valves or the exhaust valves and sometimes only on the end cylinders to accommodate variations in runner length. A racer might tighten the lash on the exhaust side to increase the exhaust event if he feels that the engine is exhaust limited. Or he might tighten the lash on the outer four corner cylinders to compensate for the longer intake runners on those cylinders. That's equivalent to running a bigger cam on those cylinders.

You may recall from Chapter 8 that sometimes you can effect dual torque peaks and a broader torque curve by running different-size (c/s area) primary pipes on alternating cylinders in the firing order. This is a fine tuning measure, but in some cases you can combine this with valve lash adjustments on selected cylinders to further tune the torque output at different speeds. In theory this is predictable, but in practice it often requires dyno verification to quantify gains.

Valve lash changes should be limited to a maximum of 0.004 inch, and consideration should be given to the known valve-to-piston clearance before going too far on the exhaust side. These tuning measures can net small gains, but the correct combination can effectively broaden a torque curve with surprisingly good results. This may be just enough to give you some added leverage on the competition without having to make major engine modifications.

Finding TDC

Locating TDC accurately is absolutely essential to proper camshaft installation. Exact TDC is the timing basis for all camshaft timing events. The method for locating it varies according to the engine's state of assembly. Whatever that is, a temporary piston stop is used to stop the piston at some arbitrary distance before and after TDC.

For fully assembled engines that are not already equipped with an accurately set TDC indicator, a threaded piston stop can be installed in the spark plug hole of the number-1 cylinder. Note that on most V-8 engines, the number-1 cylinder is almost always the farthest one forward in the V configuration. Paired rod and piston assemblies on each crank throw dictate that one is always offset farther forward than its counterpart. Study the front of the block to see which of the front cylinders is farther forward. That will be number-1.

If the degree process is being performed during engine assembly, it is best to do it with only the number-1 piston and rod assembly installed on the crankshaft. Rotating the engine to degree the cam is much easier this way. In this case, a flat bar piston stop is bolted to the block deck surface above the number one piston. This type of piston stop has a center bolt that can be adjusted to stop the piston at any desired point below TDC.

• Begin by installing the degree wheel on the crank snout, or the balancer if it is already installed.

- Before installing the piston stop, rotate the engine until the piston top visually appears to be at TDC. You should be able to see this through the spark plug hole on an assembled engine. It doesn't have to be exact—just close.

- Install a temporary wire pointer and adjust it so the tip is close to the graduated marks on the degree wheel.

- Adjust the degree wheel so the pointer indicates TDC (0 degrees) and snug it lightly.

- Rotate the engine counterclockwise approximately one-half turn and install the piston stop.

- Tighten it securely so it won't move when the piston contacts it.

- Slowly rotate the engine clockwise until the piston contacts the piston stop.

Note the pointer reading on the degree wheel. It will be in degrees before top dead center (BTDC). Record that number and then rotate the engine in the opposite direction (counterclockwise) until it completes a revolution and contacts the piston again.

- Record the reading on the degree wheel and note that it indicates degrees after top dead center (ATDC).

If your calibrated eyeball is very accurate, the recorded numbers indicate the same number of degrees on either side of TDC and the pointer reads zero with the piston stop removed and the piston brought to the top. In practice, most of us aren't that accurate, so we have to locate TDC based on a common reference point on either side of TDC. That's the piston stop. The reason you can't accurately locate TDC visually is because the piston experiences a brief period of dwell (stationary) at the top of its stroke as the rod angle transitions from one side to the other. The piston is stopped at this point and you have to split the dwell point exactly to find true TDC.

Since the piston stop does not move, it represents a fixed reference point before and after TDC. True TDC is found by splitting the difference between the degree wheel readings.

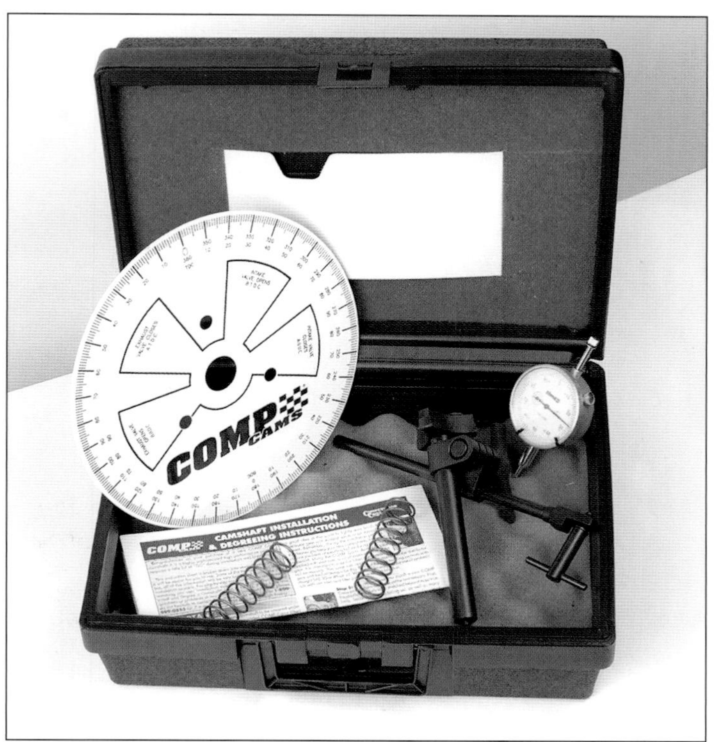

A cam checking kit like this one from Comp Cams provides all the necessary measuring tools to successfully degree your high-performance camshaft.

For example, let's say your recorded numbers are 34-degrees BTDC and 30-degrees ATDC. The exact number will depend on the depth of your piston stop in the cylinder bore, but it is all relative. TDC is halfway between the recorded readings. Loosen the degree wheel and rotate the degree wheel only until the pointer reads 32-degrees. Lock down the degree wheel and make sure not to touch or move the pointer from this point forward. Check your work by rotating the engine back and forth to the piston stop in both directions. The pointer reading should be the same in both directions (32 degrees in our example). If it is not the same, repeat the steps until the pointer indicates the exact same number of degrees before and after TDC. Once it does, remove the piston stop and degree the cam with confidence that you are locating your timing events based on exact TDC.

Degreeing the Cam

There are two methods for degreeing a camshaft. One compares the opening and closing points of the intake valve to see if they match the manufacturer's specs on the

Most degree wheels are precise if you are careful about locating the position of the pointer. You can degree accurately with all of them, but many tuners prefer the larger-diameter professional degree wheel.

The professional degree wheel offers larger spacing between degree increments for more precise positioning and it can be used to turn the engine by hand if only the number-1 piston is installed.

cam card. The other method locates the intake lobe centerline relative to TDC. Both methods are successful, but the intake centerline method does not verify the intake opening and closing points according to the cam card. Both methods are described below, but the intake opening and closing method is recommended for initial setup. Then you can check your work with the intake centerline method. In either case you need an accurate means of reading lifter travel.

I prefer the cam checking tool available from Jegs, Summit, and many other suppliers, but successful results can be obtained using a solid lifter or a modified hydraulic lifter with the internal plunger reversed to give the dial indicator plunger a flat surface to bear against. You can also locate the plunger against the edge of the lifter. Make sure that the contact is stable and that the direction of the indicator travel is parallel to lifter travel. Then adjust the dial indicator to ensure that it has enough available range to read total intake lifter travel for the number-1 cylinder.

Intake Opening Method

Install the cam with the timing marks correctly aligned for your engine. Set up your dial indicator and check lifter, or the cam checking tool in the number-1

intake lifter hole as described above. Zero the dial indicator and rotate the engine in the normal direction of rotation for several revolutions to verify that the dial indicator reads full lifter travel and returns to zero each time. You can take this opportunity to verify that lifter travel matches the indicated lobe lift on the cam card. If the lifter does not return to zero on the base circle, determine the cause and correct before continuing.

Once you're satisfied, begin with the lifter on the base circle and slowly rotate the engine clockwise until the indicator shows 0.050-inch lifter travel. Note the reading on the degree wheel. It should match the intake opening point (IO) indicated on the cam card for 0.050-inch lift. Continue rotating the engine through full lifter travel and down the other side of the lobe until you reach 0.050-inch lift before the intake closing point.

Since you know the lobe lift and the recommended closing point from the cam card, you should be able to anticipate the closing point as you rotate the engine. If you miss it, simply back up about 60 degrees to compensate for timing chain slack and approach the 0.050-inch closing point again. Compare it to the cam card and then continue rotating to verify that the lifter returns to zero again.

Your readings should show the intake opening and closing points and the total lifter travel or lobe lift. If the

Good degree wheels are marked to indicate a range where valve events normally occur. If you're measuring an event that occurs outside that range, it is probably incorrect and you should recheck your work.

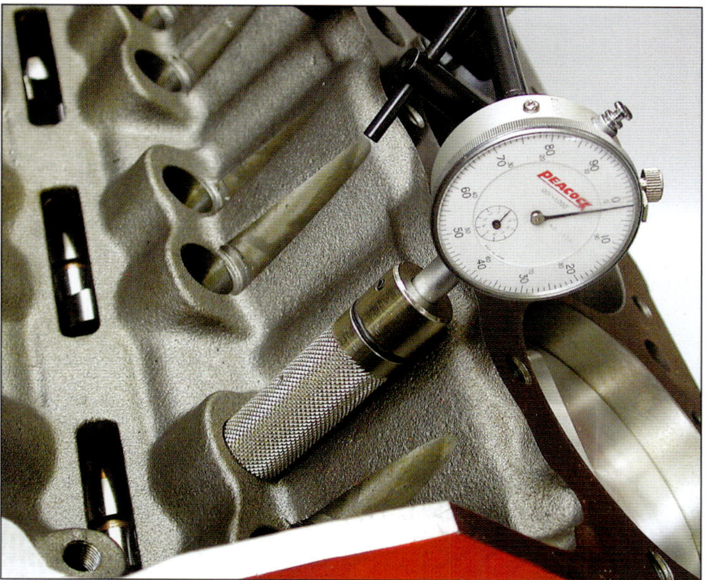

A cam checking tool reads lifter motion directly without the need for dial indicator spindle extensions. It can only be used with the cylinder heads off the engine, but it is the most accurate method.

intake opening event doesn't match the cam card, you will have to advance or retard the cam to bring it into spec.

For example, if your cam is supposed to open the intake valve at 36-degrees BTDC and close at 70-degrees ATDC (at 0.050-inch lift), but your measurements show that it is opening 34-degrees BTDC and closing 72-degrees ATDC, the cam is retarded. The valve event is occurring later than the recommended spec. If it were to open at 38-degrees BTDC and close at 68-degrees ATDC it would be 2-degrees advanced because the valve event is occurring 2 degrees earlier than specified.

In either case it is easy to correct using offset cam bushings or a crank gear with multiple keyways. Both allow you to adjust the position of the cam and then recheck it for compliance with the cam card specs. Note that they can also be used to reposition the cam if you deliberately choose to advance the cam to promote low-end torque or retard the cam for a little more top end power.

If your degree results are plus or minus 1 degree of your published specs, consider leaving the engine as assembled because it is entirely possible that the small degree wheel you are probably using is not that accurate. Larger-diameter degree wheels space the degree marks farther apart and, therefore, have a greater chance of improved accuracy.

You can check the accuracy or your wheel by placing it on a large sheet of paper and marking the four 90-degree positions of the wheel. Then move the wheel to various positions and check to see that each 90-degree mark is an equal number of degrees from 90. You may well find that your wheel is not completely accurate. This is why fussing over less than 2 degrees (unless for example, the cam is retarded 2 degrees and you want 2 degrees advanced) may not be worth the effort.

Intake Centerline Method

The intake centerline method finds the location of the intake lobe centerline relative to TDC. The recommended intake centerline is indicated on the cam card, and when correct it should yield the specified intake opening and closing points when you degree the cam. Finding the centerline is easy.

Rotate the engine clockwise until you find the maximum lobe lift, then zero the indicator. Now rotate backward about 0.100 to 0.150-inch to compensate for timing chain slack. Then rotate clockwise until you reach 0.050 inch. This is 0.050 inch before max lift.

Note the reading on the degree wheel. Then continue over the nose of the cam until you reach 0.050 inch again. This is the 0.050 inch after max lift. Note the

Rocker Ratio Tuning

You can tune a mechanical or solid lifter cam to some degree with small valve-lash adjustments, but you can tune any pushrod-style cam with a rocker arm change. For example, a Chevy 1.5:1 rocker arm can be replaced with a 1.6:1 rocker to achieve a 10-percent gain in rocker ratio. That doesn't mean a 10-percent increase in lift, but the effective increase is substantial and often quite useful. On a Chevy this increases valve lift by about 0.030-inch. As a general rule, the higher the lobe lift, the greater the increase. To be certain of your gain, multiply your lobe lift by the new rocker arm ratio.

Net Lift = lobe lift x new rocker ratio

For example, a lobe lift of 0.350-inch provides a net lift of 0.525 inch except with a mechanical cam. A 1.65:1 rocker arm bumps that value by more than 0.050 to 0.577 inch; a substantial increase. Equally important are the other effects a ratio change brings to the table.

A higher-ratio rocker arm accelerates valve opening and closing events at a faster rate, effectively increasing duration with the same opening and closing points. The general rule is one degree of duration for every half point of rocker ratio increase. Hence a switch from 1.5 to 1.6:1 can net a 2-degree increase in duration. Power gains acquired through ratio changes are split between the increase in lift and the increase in duration.

Be mindful that a rocker ratio increase may affect valve to piston clearance, valve spring coil bind, and retainer to seal clearance, and the faster valve action may induce valve float in some cases. Generally these problems don't occur, but you have to check for them and take appropriate action if necessary.

Ratio changes are often made on either the intake or the exhaust only to help crutch a deficiency elsewhere, and sometimes tuners run more rocker ratio on the end cylinders to compensate for differences in runner length. Whatever the case, rocker ratio is an effective tuning tool for racers who understand the advantages it can provide.

You can read net valve lift and duration with the valvetrain installed by checking it at the retainer. This method incorporates the rocker ratio and any clearance lash to yield an accurate indication of true valve motion.

degree wheel reading again. Now add the two readings together and divide by 2 to find the center line. It should match the cam card.

For example, if your numbers are 80 and 132:

(80 + 132) ÷ 2 = 106-degree centerline

The cam card indicates the correct installed intake centerline. If it calls for 106 degrees and you come up with 108 degrees, the cam is early and you have to retard it 2 degrees to bring it into spec. If you get 104 degrees the cam is retarded and you have to advance it 2 degrees to correct it.

If you have degreed the cam with the intake centerline method, go back and check to see if the intake opening and closing points match those indicated on the cam card. If incorrect, determine the direction of error and reposition the cam accordingly.

Advertised Duration vs. 0.050-inch Lift Duration

The difference between advertised duration and duration at 0.050-inch lift doesn't have to be confusing. All cam manufacturers select an opening and closing lift point to specify the duration of their cams. This is typically a point of discernable lifter motion such as 0.006 inch for COMP Cams or 0.004 for Crane Cams. These lift points are the basis for the advertised duration that all cam manufacturers use. Advertised duration is problematic because manufacturers use different lift points to specify duration, thus making cam comparisons difficult.

Years ago all manufacturers finally settled on a universal checking point of 0.050-inch to compare cam specs from different companies. This spec was adopted by the SAE and is the standard checking point for all camshafts.

Later, Harvey Crane introduced the concept of hydraulic intensity, which is the time in crankshaft degrees that it takes for the lifter to move from its advertised duration checking point to the universal 0.050-inch lift point. It's determined by subtracting the duration at 0.050 inch from the advertised duration. The smaller the number, the greater the hydraulic intensity, which indicates a more aggressive cam. Naturally comparisons can only be made between cams with the same advertised checking point relative to the 0.050-inch standard. Cams with a more aggressive hydraulic intensity typically exhibit more valvetrain noise due to the steep lift curve, but they generally deliver more performance without sacrificing idle quality.

Calculating Valve Overlap

Overlap is the number of degrees where both valves are off their seats at the same time. It is a combination of the intake opening event and the exhaust closing event. Adding these two points together yields valve overlap.

$$\text{Valve Overlap} = IO + EC$$

For example, a cam with an intake opening point of 29-degrees BTDC and an exhaust closing point of 23-degrees ATDC has a valve overlap of 52 degrees.

$$29 \text{ degrees} + 23 \text{ degrees} = 52 \text{ degrees overlap}$$

Camshaft Formulas at a Glance

Duration at Checking Lift = opening point + 180 + closing

Intake Centerline ATDC = (intake duration ÷ 2) − IO

Exhaust centerline BTDC = (exhaust duration ÷ 2) − EC

LSA = (intake centerline + exhaust centerline) ÷ 2

Net Lift (solid lifter) = (lobe lift x rocker ratio) − valve lash

Net Lift (hyd. lifter) = lobe lift x rocker ratio

Intake Centerline BTDC = (degrees at 0.050 before max lift + degrees at 0.050 after) ÷ 2

TOOLS AND EQUIPMENT

Many of the formulas in this book require precision measurements to ensure accuracy. This chapter familiarizes you with the tools required to gather these measurements and some strategies for organizing and storing them for future reference. Most of the tools presented here are relatively inexpensive and can be added to your engine building inventory one or two at a time. Many of the hard-measurement tools shown are sourced from summitracing.com and are indicated where applicable. Others come from a variety of manufacturers who build specialized tools for unique tasks. The important thing is that all of these tools are very easy to use and they deliver the precision results necessary for competent engine blueprinting and assembly.

Accordingly, it is important that you maintain good records of your measurements. A sample engine assembly sheet is included on page 145 for you to copy. It contains entry spaces for all the critical measurements you will normally require.

Over the course of your engine building experience you also may determine that there are other bits of data that you wish to keep track of for future reference. In much the same ways as explained in Chapter 13, you can create your own blank build (assembly) sheets for recording various measurements and arrange them any way you wish. The sample shown on page 145 could be made by creating a table in Microsoft Word or in Excel. You can easily rebuild this one in your own computer and modify or rearrange it to suit your personal needs. Note the secondary content sheet (page 144) that allows you to keep track of your engine's contents including parts, part numbers, sources, and contacts.

You can ultimately build other sheets to store your engine balance data, cam cards, cam setup information, or anything else you deem important. The spreadsheet is the easiest way to do this, and in some cases you might even include a formula that automatically calculates something from measurements that you have entered on that page. You can make your entries directly in the PC or print out blank sheets to be filled out as required. Use the sample page as a reference or build your own. Measurements are a core component of any good engine build. You can never go wrong if you have accurate records. And if you are brainstorming engine combos on simulation software, you can open a second window and enter parts and specs into your engine build content sheet at the same time.

Measurement Tools, Standards and Accuracy

The most important characteristics of all measurement tools are accuracy and consistency. Modern measurement tools are exceptionally accurate and, when properly cared for, they are remarkably consistent. Most instruments that measure in thousandths of an inch will agree with each other on any given day as long as the temperature of the instrument and the component being measured are close to similar. Repeatability is a vital component of any measuring device and part of that responsibility falls on the measurement technician. The device must be used the same way every time and the component temperatures must be equivalent.

Suppose you live on the East Coast and the temperature outside is 10 degrees F on a crisp January morning. You can't bring in a crank that has just made the trip to your location in the back of an open pickup truck and expect the measurements you take to provide an accurate readout of bearing clearances in a bare block with bearings that have been sitting in your heated assembly room all night. When you start measuring in tenths of a thousandth of an inch, temperature becomes a critical element, hence the use of checking standards to ensure accuracy. If you ask a machine shop to fit a set of pistons to a block with 0.0035-inch piston-to-wall clearance they will almost always hit it right on the nose. But if you get it back to your shop or home garage and your checking equipment indicates that the pistons are fitted at 0.004 to 0.0045 inch and a couple are even nudging 0.005 inch, whose measurement is correct?

Why is this so critical? Well, it's because engine components are specifically designed to work together with clearances that account for variations in engine operating temperature, lubrication viscosity, engine speed, loading, and severity of operation. Lubrication clearances or tolerances are the facilitators that allow complex pieces of machinery like automotive engines to run smoothly and provide long-term durable service. Some of the instruments you may use measure other important things such as pressure, volume, weight, movement, and so on. These instruments are also influenced by temperature and the way you use and care for them.

All of the measurement tools presented below are designed to provide accurate measurements that will adhere to common standards. As long as you keep them clean and take proper care of them, they will deliver consistently accurate measurements that you can accept with confidence.

16-inch Pro Degree Wheel

Larger-diameter degree wheels are popular with most professional engine builders because they afford greater precision and ease of use. They are sturdy enough to use for rotating the engine during cam setup when only the crank, camshaft, and one piston are installed. These wheels offer full 360-degree bright markings in 1-degree increments and are made from anodized aluminum for durability.

A larger-diameter degree wheel typically yields more precise measurements. Degree markings are spaced farther apart and they are easier to read.

32-inch Precision Metal Straightedge

An inexpensive metal straightedge is used to check for straightness and alignment on engine blocks and cylinder heads. With accuracy to within 0.001 inch, it is used in conjunction with feeler gauges to check for straightness or high and low spots. It is particularly useful for checking the straightness of cylinder block and head deck surfaces and the alignment of main bearing housing bores.

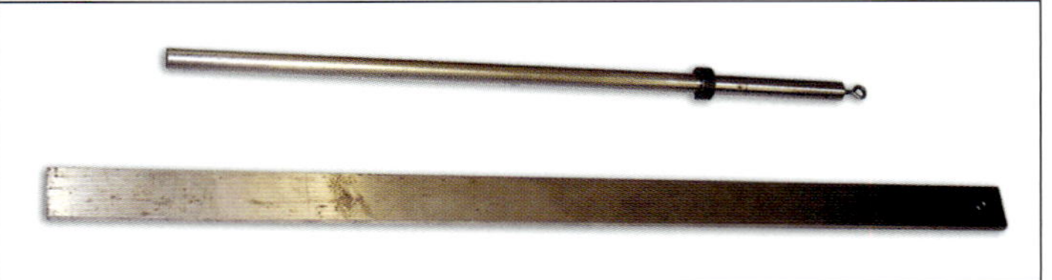

Check cylinder block deck surface straightness with a precision metal straightedge. Align it from corner to corner and use feeler gauges to check for low spots. You can also check cylinder head decks for straightness and warping.

Burette and Stand Kit

Used for cc'ing intake ports, combustion chambers and other volumes. It comes with a clear molded glass burette with a Teflon petcock. The burette is graduated in 0.20-cc increments and sharp easy-to-read lines that are numbered every 2 cc's. A kit includes an adjustable metal

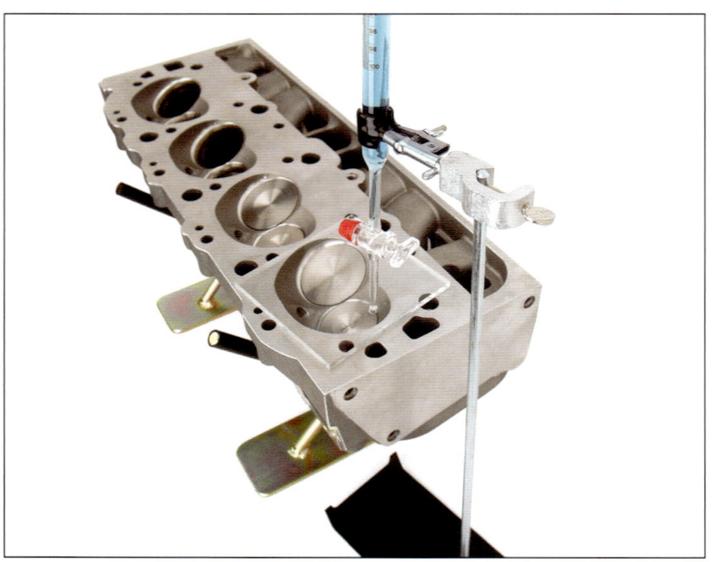

A graduated burette kit with stand is essential for cc'ing intake ports, combustion chambers, and other small volumes.

stand for easy positioning with any size cylinder head or any other checking requirement.

Cam Checking Tool

A cam checking tool slips into the lifter bore with a cam follower and dial indicator for accurate measurement of lobe lift and base circle run out. It can be used in conjunction with a degree wheel for cam checking and degreeing. It comes with two followers: one for flat tappets and one for roller lifters. The double-ended design allows it to fit both GM (0.842-inch) or Ford (0.875-inch) lifter bores. It comes complete with dial indicator, locking set screws, and rubber O-rings to hold it stable in the lifter bore.

The cam checking tool fits snugly into the lifter bore to provide a direct readout of camshaft lobe lift. It's the best way to degree a camshaft in a short block or with the heads installed and a dial indicator with a spindle extension.

Degree Wheel Kit

A cam checking and degree kit that can be used with the heads on or off the engine. Includes a cam lobe checking fixture with a 1-inch dial indicator and a 5-inch extension, TDC locators, and an 11-inch precision degree wheel. It can be used on blocks with 7/16- or 1/2-inch head bolt holes or on cylinder heads with 1/4- or 5/16-inch valve-cover bolt holes. The degree wheel is separately

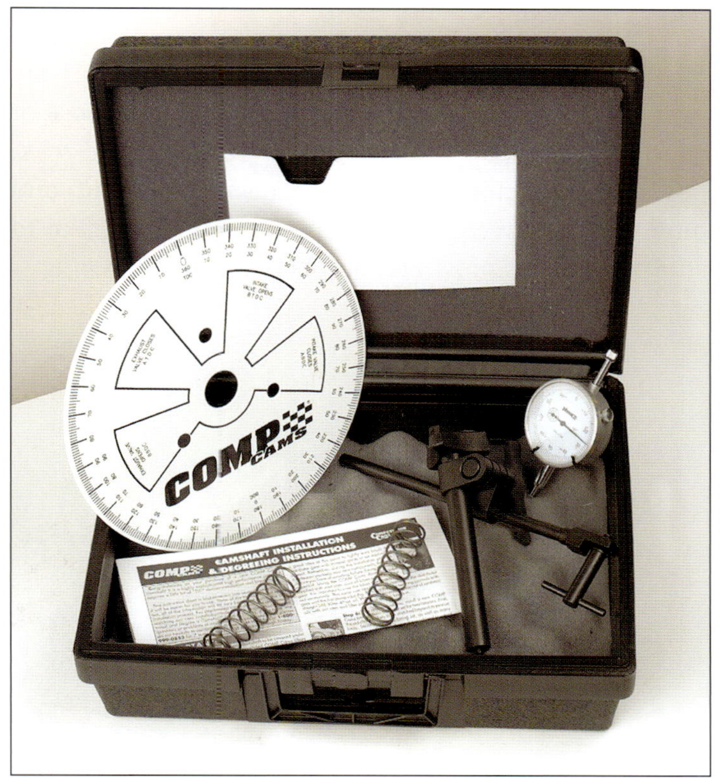

A degree wheel kit contains all the measuring tools required to accurately degree a camshaft.

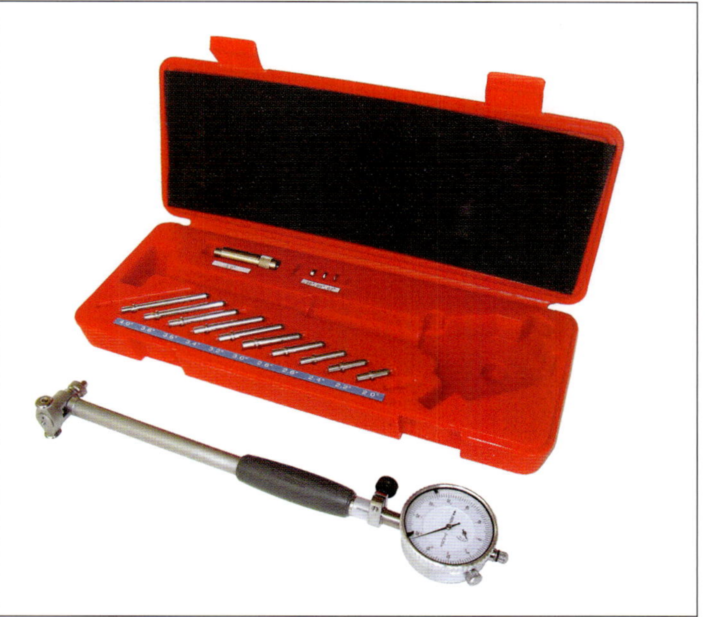

Use a dial bore gauge to check cylinder bore diameter and taper, rod and main bearing diameter, and bearing housing bore diameters. Dial bore gauges read to a tolerance of 0.001 inch.

highlighted for intake and exhaust events, intake centerline, and exhaust centerline in 1 degree increments.

Depth Micrometer

A depth mic is handy for quick measurements of deck height, piston rock, valve depth in combustion chamber, and other measurements that require a depth probe. Most tools have a range of zero to 1/2 inch or zero to 1 inch in 0.001-inch increments, just like a micrometer.

Dial Bore Gauge

A dial bore gauge is the only accurate way to measure cylinder bores, main bearings, rod bearings, and other inside diameters. They take precise measurements to within 0.0005 inch and are used for checking cylinder bore straightness and taper and for determining piston to wall clearance. Most units come with a full range of add-on anvils for measuring bore diameter from 2 to 6 inches and up to 6 inches deep.

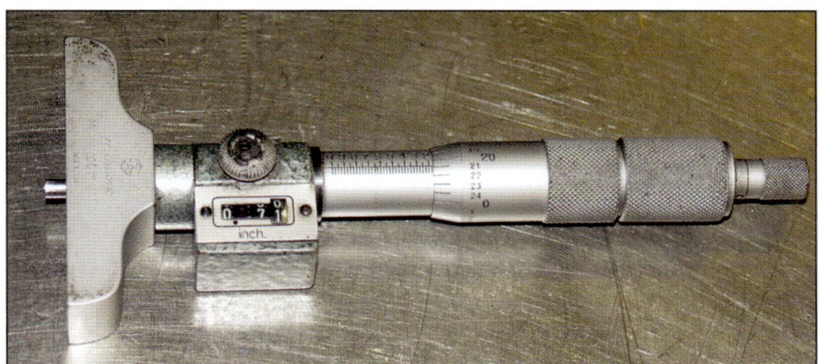

A depth micrometer makes quick work of checking piston deck height and piston rock. They're also useful for checking the depth of blind head bolt holes and other depth measurements.

A digital dial caliper takes quick inside and outside measurements to 0.001 inch. This is another essential tool that should be in every engine builder's toolbox.

Digital Caliper

Perhaps the engine builder's second most handy tool. The digital stainless steel caliper has a large LCD display window with direct readout in metric or SAE units. It measures from 0 to 6 inches in 0.001-inch increments. Most units come with a battery included and a protective storage case. Can be used to measure piston skirts, journal diameters, and assorted depths and heights with the inside or outside jaws.

Dial Indicator

This is the engine builder's basic measuring tool. Used in conjunction with a magnetic base, bridge stand, or other holding devices, dial indicators are used for checking crankshaft thrust, camshaft end play, deck height, valve travel, and many other measurements from zero to 1 inch in 0.001-inch increments. They have a rotational clamp for zeroing the indicator at any point, interchangeable contact points, and continuous dial gradation in 0.100 inch per dial rotation with a smaller dial indicating the number of total dial rotations up to 10. Special extended-travel versions are available for checking stroke length and other longer measurements.

Dial indicators provide a direct readout of precision measurements on an easy-to-read dial with 0.001-inch gradations. They can be zeroed in any position for quick, easy measurements.

Dial Indicator Bridge Stand

A bridge stand positions a dial indicator directly over a deck surface or cylinder bore for the purpose of checking deck height, valve pocket depth, piston rock, piston-to-head clearance, deck flatness and runout, and other comparative measurements of one flat surface to another. Accepts most standard dial indicators.

Use a dial indicator on a bridge stand to measure differences in parallel surfaces such as piston tops and cylinder block deck surfaces.

Dial Indicator Magnetic Base

A magnetic dial indicator base allows you to position a dial indicator in almost any position necessary to obtain a precision measurement. They accept all standard dial indicators and feature a heavy-duty magnetic base for easy attachment to cylinder blocks and iron heads. The base features a powerful magnet with on/off release lever and integral clamps for horizontal or vertical adjustments. Some base styles have a semi-rigid articulating arm that can be twisted to almost any position. The dial indicator itself is not included in most magnetic base kits.

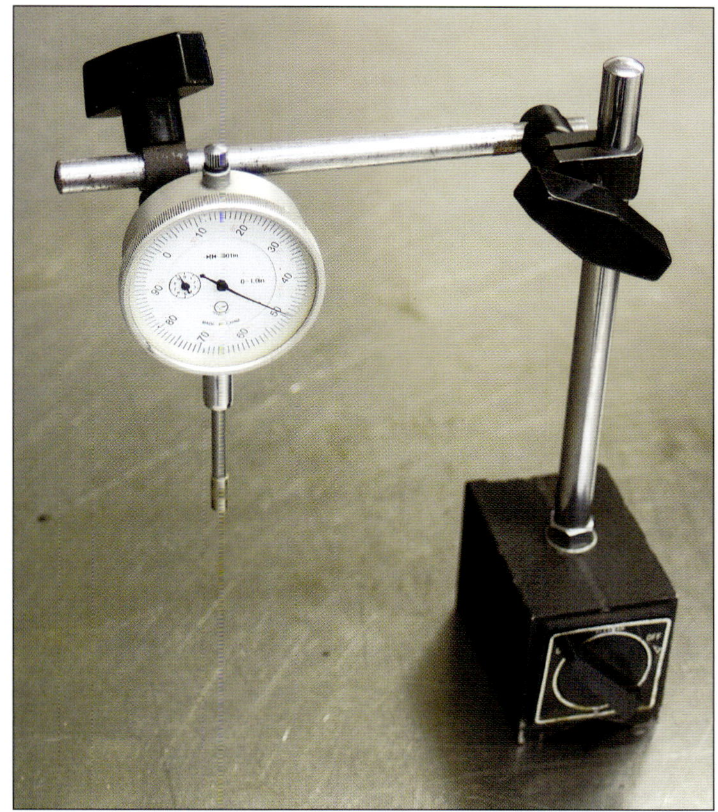

A magnetic base is used to mount dial indicators in a variety of positions to facilitate precision measurements.

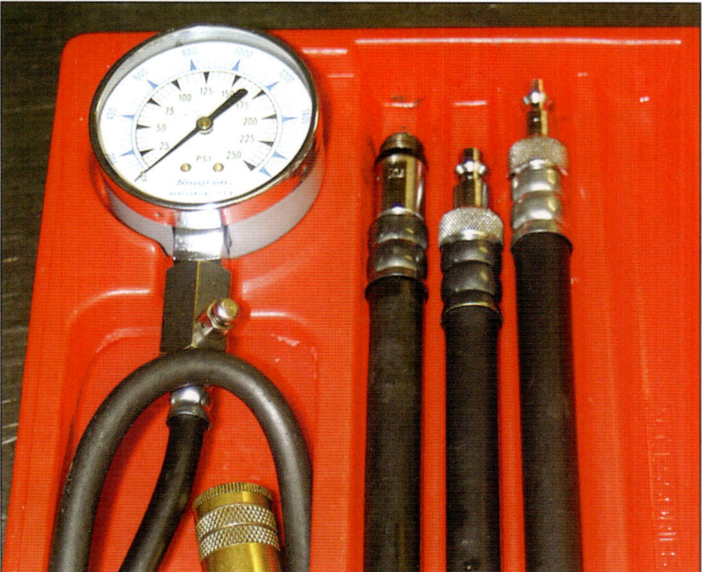

Cranking compression is measured in each cylinder while the engine is cranked with the ignition off and the throttle held open. All cylinders should read within 10 percent of each other.

Digital Compression Gauge

A digital compression gauge reads engine cranking pressure in individual cylinders up to 300 psi. A good one comes with common spark plug hole adapters and includes automatic shutoff once it has recorded an average pressure. It features zeroing with a convenient push-button relief valve. A sound engine will have compression readings within 10 percent of each cylinder, but the actual amount of compression varies according to contributing factors such as cam timing and static compression ratio. Compression testers are primarily used to assess the condition of valves and piston rings in each cylinder.

Digital Torque Adapter

PowerBuilt Tools by Alltrade makes a great digital torque-wrench-calibration tool that calibrates any torque wrench and doubles as an actual torque wrench when attached to a standard ratchet or breaker bar. It also provides visual and audio indicators when reaching specified torque and records the last 50 torque-value readings. It calibrates from 29.5 to 147. 6 ft-lbs.

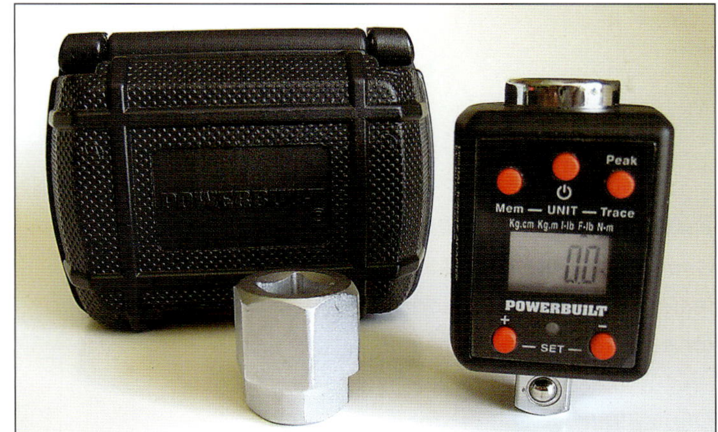

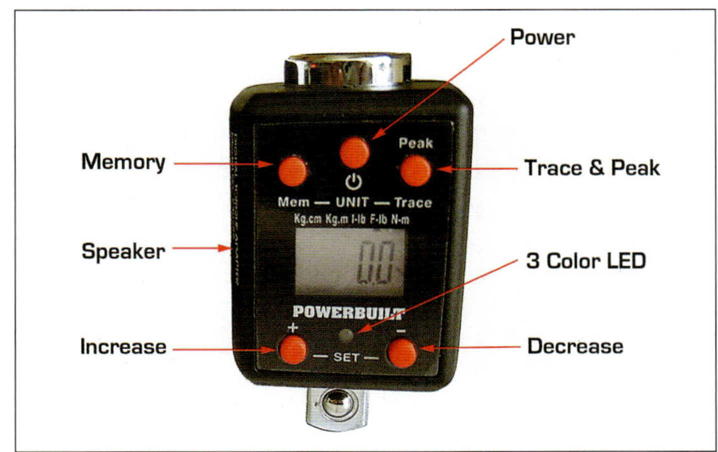

Keep your torque wrench in perfect calibration with Power-built's inexpensive digital torque converter.

One of the most useful tools is a valvespring pressure tester with a digital readout and adjustable positive stop for easy repeat measurements.

Digital Valvespring Tester

A digital or manual valvespring tester is used to check spring pressures at the manufacturer's recommended installed height. By checking tension at different heights, the spring rate can be calculated and compared to the manufacturer's published specs. Most units feature a positive stop at an adjustable desired height. Digital units are convenient because you don't have to read small gradations on a dial.

Feeler Gauges

Feeler gauges are typically used for measuring connecting rod side clearance, ring groove clearance and end gap, valve lash, and other measurements that require a flat blade of calibrated thickness. They are available at any

Feeler gauges are the most convenient way to measure small clearances such as valve lash, connecting rod side clearance, or piston ring gap.

hardware store and most come in a range from 0.002 to 0.035 inch. Those intended for valve lash work often have angled tips for easier insertion between the valve stem and rocker arm tip and they come with fewer sizes that accommodate the most commonly used valve lash settings.

Light Checking Springs

Light tension springs are used to hold the valves in place for flow bench work and other tasks such as measuring valve lift and valve-to-piston clearance, degreeing a cam, or checking rocker arm ratios. They also hold the valves closed with sufficient tension for cc'ing combustion chambers and ports or while adjusting valve open position on a flow bench.

Light tension checking springs allow you to assemble valves in a cylinder head to check valve-to-piston clearance without having to compress the heavier valvesprings. They are also handy for holding the valves closed while cc'ing ports and combustion chambers.

Mr. Gasket's Hot Rod Calc

Another great tool from Summit Racing is the hand-held Hot Rod Calc; a convenient hand calculator containing all the formulas in this book and more, including vehicle dynamics calculations, weather and performance predictions. It is probably one of the first things you should buy when you start building your own performance-engine tool kit.

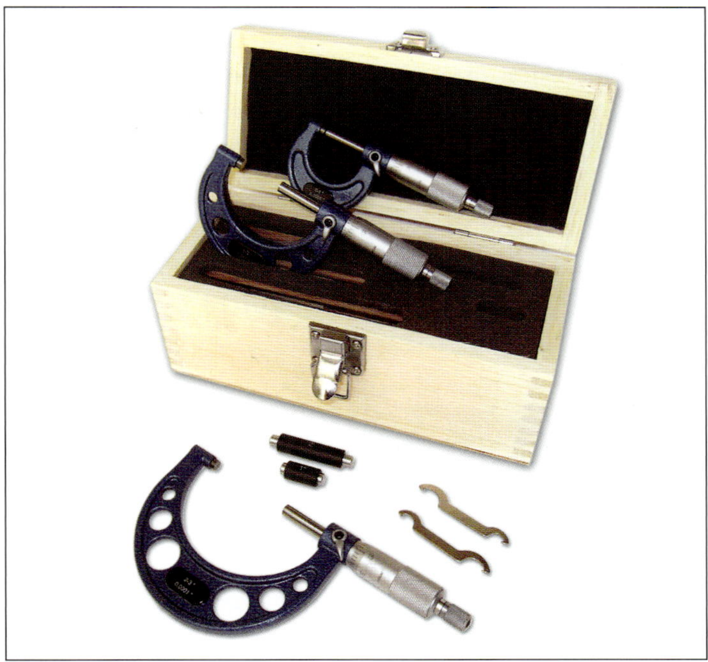

Outside micrometers are essential for verifying crankpin journal diameters and piston skirt clearances. With a little practice it's easy to acquire a good feel for precision mic measurements.

PerformAIRE Air Quality Computer

This handheld digital unit measures density altitude to within 50 feet. Essential for calculating jet changes or calculating fuel pressure changes on EFI systems. It also monitors temperature, barometric pressure, relative humidity, and other values that are displayed on an easy-to-read LCD screen.

Mr. Gasket's Hot Rod Calc calculator, available through Summit Racing is the ideal companion for racers and engine builders needing to make quick calculations.

Outside Diameter Micrometer Set

A set usually consists of three separate micrometers capable of measuring from zero to 1 inch, 1 to 2 inches, and 2 to 3 inches. Most have a positive locking clamp and a sensitive ratchet stop to prevent overtightening. Micrometer sets come in a wooden case and include calibration blocks so you can zero them correctly. They are primarily used for measuring bearing journals, wrist pins, and other rounded surfaces. If you intend to check piston skirts, you will have to purchase separate larger micrometers in the 3- to 4-inch and 4- to 5-inch ranges.

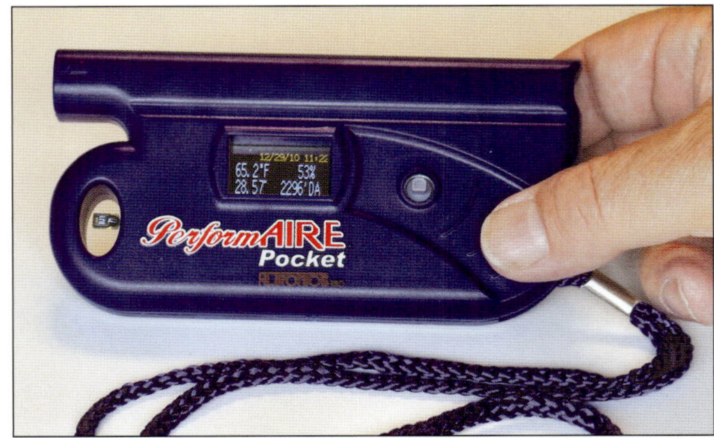

New PerformAIRE Pocket Weather Station from Altronics provides direct readout of track conditions including barometric pressure, temperature, humidity and current density altitude for local atmospheric conditions.

Pushrod Length Checkers

COMP Cams makes the best pushrod length checkers because they have graduated vernier caliper-like scales marked directly on the adjustable pushrod. Each full turn advances the length of the pushrod by 0.050 inch so all you have to do is count the number of turns from the fully closed positions. Use them to establish proper rocker arm geometry before ordering exact-length pushrods.

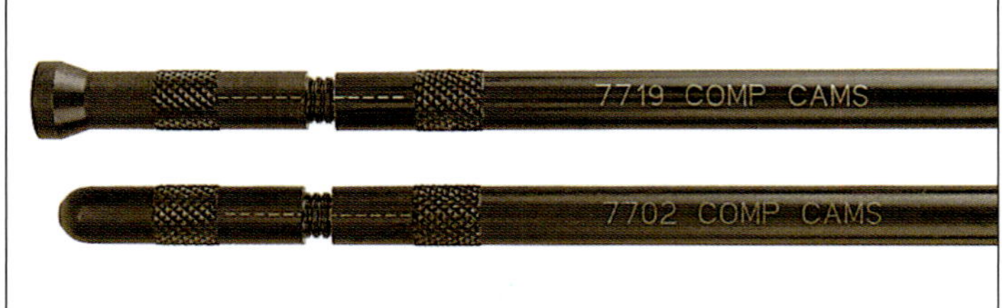

Adjustable pushrods are perfect for determining the ideal pushrod length. Dial in your rocker arm geometry, then order pushrods to the exact indicated length.

Rod Bolt Stretch Checker

A dedicated tool for accurately measuring rod bolt stretch with a dial indicator. Compares the length of unstretched bolts to the length after torquing. Allows the bolt to be further tightened to manufacturer's recommended stretch rather than a specific torque.

Torque rod bolts to recommended specs, then stretch them to the manufacturer's recommended stretch by tightening them farther and checking them with a rod bolt stretch gauge.

Snap Gauge

Snap gauges are handy adjustable gauges for measuring the inside width or diameter of various ports or openings. They have a compressed spring that provides tension to expand against the sides of the object being measured. A locking device freezes the measurement, which is then read using an outside mic. They are available in a common range of sizes or in a kit. You won't use them a lot, but they are handy when necessary. They are inexpensive and available at most local tool outlets.

Checking port dimensions is easy with the appropriate size snap gauge. An inside micrometer is used to read the final height and width measurements once the snap gauge has been locked.

Stroke Length Checker

A precision stroke length checker consists of a pair of equal-length aluminum bridge stands with V-cuts on the bottom to center the device on adjacent main bearing journals. The gap between them is bridged by a bar with

The basic stroke checking tool is relatively inexpensive and easy to use for pro and amateur engine builders. This one is available from Powerhouse.

a dial indicator and extension that reads the stroke as the crank is rotated in a set of V-blocks. It requires a special dial indicator with a capability of reading up to 4½ inches of travel.

Summit Cam Checking Fixture

Summit's heads-off cam checking fixture positions the dial indicator and 5-inch extension in the correct position over the lifters so you can obtain accurate readings without the need for a magnetic dial indicator base.

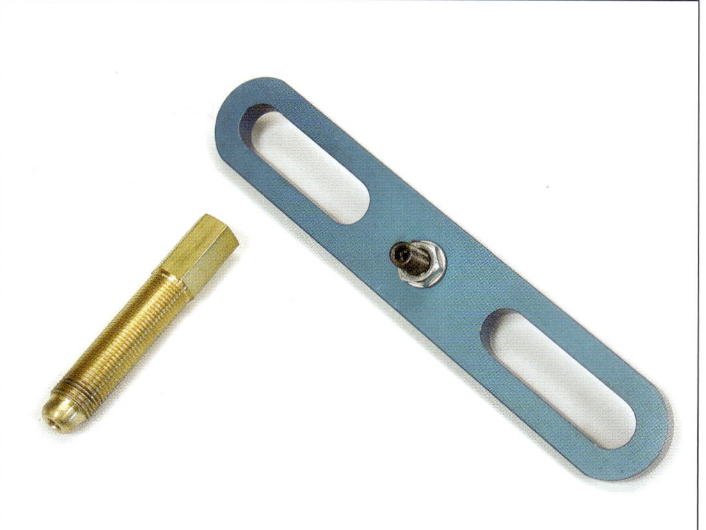

A TDC indicator piston stop is used to align a pointer with the degree wheel for indicating the exact position of TDC (top dead center).

A heads-off cam checking fixture makes quick work of checking the cam position when you are working at the short block level.

TDC Indicator Stop

A universal deck mounted bar with an adjustable positive top for finding TDC in conjunction with a degree wheel. Mounts to the block deck surface using existing head bolt holes.

Valvespring Height Micrometer

These come in two sizes and allow you to directly measure valvespring installed height with the correct retainer. They measure within 0.005 inch and have a height range from 1.400 to 1.800 inches or 1.800 to 2.200 inches. A later version is also available for checking the newer beehive-style springs. They are the most accurate way of checking installed spring height.

A valvespring height micrometer checks actual valvespring installed height with the correct valvespring retainer. It provides a direct readout on a micrometer-like scale.

HOW TO BUILD AN ENGINE MATH SPREADSHEET

	A	B	C	D	E
2	Bore	4.00	cid	349.85	B2*B2*B3*.7854*B4
3	Stroke	3.48	cyl. vol.	43.73	B2*B2*B3*.7854
4	Cylinders	8			
5	RPM	6000	P/S ft/min	3480	B3*B5/6
6					
7					
8	Displ.	327	Bore ?	4.00	sqrt(B6/(B7*.7854*B4))
9	Stroke	3.25			
10					
11	Displ.	372	Stroke ?	3.48	B11/(B12^2*.7854*B4)
12	Bore	4.125			
13					
14					

Our sample calculator as constructed in Microsoft Excel. You can follow the sequence in the example using this screen shot or the accompanying printout.

Setting up a PC-based spreadsheet to hold your important engine math formulas is a handy way to keep them all together and use them to calculate anything you want to know just by inputting the appropriate values. This assumes that you already have a core level of personal computer skills, and are able to open, manipulate, and save files without distress.

A spreadsheet is nothing more than a big graphic calculator that you can easily program to calculate and analyze data. It is your own personal database, and you can configure it to make all the calculations in this book by inputting the appropriate information. It's a great tool for brainstorming engine combinations. The sample presented here is done in Microsoft Excel, which many people already own. There are also online open document spreadsheets in what's called the computing cloud. They are equally good and pretty secure. If you don't want to have to log on to do some calculations, you can use Excel or any number of free spreadsheets available on the internet for downloading.

All spreadsheet software operates the same way. When you first open a new file you find a full page of blank rectangles called cells. They are organized into numbered rows and lettered columns. If you scroll down or across the page, you'll find that it seems to go on forever. You won't need all that space, but it's nice to know it's there just in case.

Each cell can relate to any other cell or multiple cells depending on how you manipulate them. Cells can hold plain text, which can describe what is happening with data in an adjacent cell, and they can perform spectacular calculations in a heartbeat. This allows you to build small combinations of cells into stand alone calculators that you can label and tag with titles and the names of various values that are being input for calculation. You can arrange or group the cells any way you wish to personalize your own engine math calculator.

You can also put different types of formulas on different pages to keep them separate just like the chapters in this book. I'll show you how in a moment, but first you need to learn a few more things about spreadsheets.

At the top is a blank bar called the formula bar. You can input values at the formula bar or directly into the cells. When building a formula, it's best to select the appropriate cell by clicking on it and then building the formula in the formula bar. If you enter something directly into the cell, it also displays in the formula bar as long as you have that cell selected.

When you enter a formula into the formula bar it remains resident in the cell you have selected, but only displays in the formula bar. The answer that the formula calculates displays in the selected cell. Cells appear blank until you enter data in the form of text or a numerical value, or until a resident formula makes a calculation.

Formula calculations are based on data that you input to other cells that are appropriately labeled to reflect the type of data being input. A formula may involve a simple multiplication function, or it may incorporate multiple functions like many of those in this book. Each formula must be entered in an exact sequence or it won't function.

The good news is that you only have to get it right once. After that it will always work because you'll only be inputting values to other cells that the formula cell refers to for its calculation. If you can't get it to work, consult the help screen or the manual. It is usually a simple missed keystroke that prevents the correct calculation.

If you have to make adjustments to the formula, do them in the formula bar. Keep in mind that the formula usually refers to more than one cell for its input. Cells are referred to by their location. For example, the seventh cell

down from the top in column A is called A7. Another cell on row 7 might be selected under column D. That would be cell D7. If you input a formula into cell A1 that tells it to multiply the value in A7 by the value in D7 it will do so and display the answer in cell A1 as long as there are values entered in the two reference cells. The formula in A1 remains hidden. Only the answer shows.

Finding Displacement

Here's a simple example (see facing page or page 141) using the formula for engine displacement: Recall that displacement equals bore^2 x stroke x 0.7854 x the number of cylinders. Let's find the displacement for a 350 Chevy whose bore is 4.00 inches and the stroke is 3.48 inches. You can build your calculator anywhere on the page, but for now let's stick to the upper left-hand corner. Begin with row one, column A.

Step 1

Select cell A1 by clicking with your mouse and type the word "Displacement" into the cell or the formula bar. Note that it displays in cell A1 and because it is too long, it overruns the adjacent cell (B1) to make one long cell for the word displacement. You can make a title like this anywhere you want. Underneath the title we are going to name the input values for our displacement calculator. They are all in column A.

Step 2

Type the word "Bore" into cell A2. Note that any cell without an embedded formula simply displays what you type into it. It can be a name, a number, or whatever.

Step 3

Now move to cell A3 and type "Stroke."

Step 4

In cell A4 type "Cylinders" to represent the number of cylinders in your engine. The input values for these cells are entered directly opposite them.

Step 5

So right next to bore in A2, type the bore dimension "4.00" into cell B2.

Step 6

Move to cell B3 and type the stroke value "3.48." Then type the number of cylinders into cell B4. In this case it's 8. Now you see that column A has labels and column B is where you input the appropriate value for each label.

Step 7

Move to column C and type "cid" into cell C2 to represent the displacement.

Step 8

Type "Cyl. Vol." into C3 for the cylinder volume.

Now you have column A for naming the inputs, column B for entering the inputs, column C for naming the answers, and column D for displaying the answers. You have already input the appropriate bore and stroke and number of cylinders in column B. All that's left is to enter the correct formulas into D2 and D3.

Formulas usually start with the equal sign (=) followed by cell references and the necessary mathematical symbols. There's a way to input a number squared, but for now we'll do it the long way to illustrate the process:

Step 9

Select cell D2 and type the following formula into the formula bar: =B2*B2*B3*0.7854*B4. Note that the star key above the numeral 8 on your keyboard is used for the multiplication sign. In the formula you just typed, you are telling the computer to multiply the value found in B2 times itself times the value in B3 times 0.7854 times the value in B4.

Step 10

After typing the formula, hit enter and the formula will reside in cell D2. Since you have already input values for it to refer to, it calculates the displacement and displays it in D2. The answer is 349.85 ci. (Later when you are more comfortable, you can go to the menu bar on top and adjust that cell so it displays two or more decimal places or simply round off the answer.)

Step 11

Now move to cell D3 (opposite where it shows cyl. vol. in C3). Select cell D3 and type =B2*B2*B3*0.7854 into the formula bar. Note that it is the same formula, minus the value for the number of cylinders. That's because we want it to show the displacement or volume of just one cylinder in D3.

Now the beauty of the spreadsheet is that the formulas in D2 and D3 remain resident in those cells, and they recalculate and display a new answer anytime you change the input values in column B.

Step 12

To try it, select B2 for the bore. Delete the "4.00" and enter "4.03." Leave the other two entries in column B the same. This gives you the displacement for a 4.030-inch bore with the same stroke. The answer is 355.12 ci. You should see the answer in D2 change to 355.11 and the answer in D3 change to 44.39.

Now you have built a simple displacement calculator.

Finding Piston Speed

You can do the same for the compression ratio formula or another formula elsewhere on the page or on a different page. To take it a bit further, let's say you are playing around with bore and stroke combinations and you are concerned that the piston speed might be too high if you use a longer stroke. You can add that little bit of information right into your displacement calculator and see the piston speed at the same time as you view the cylinder volumes. All you need to do is add an engine speed value and a formula for piston speed.

Step 1

Move down to cell A5 and type "RPM."

Step 2

Then, move over to cell C5 and type "PS ft/min" for the piston speed in feet per minute.

Step 3

Enter the formula for piston speed into cell D5 using the formula bar.

$$\text{Piston Speed in ft/min} = \text{stroke} \times \text{RPM} \div 6$$

So you enter: = B3*B5/6

Excel Spreadsheet Calculator

	A	B	C	D	E
2	Bore	4.00	cid	349.85	B2*B2*B3*.7854*B4
3	Stroke	3.48	cyl. vol.	43.73	B2*B2*B3*.7854
4	Cylinders	8			
5	RPM	6000	P/S ft/min	3480	B3*B5/6
6					
7					
8	Displ.	327	Bore ?	4.00	sqrt(B6/(B7*.7854*B4))
9	Stroke	3.25			
10					
11	Displ.	372	Stroke ?	3.48	B11/(B12^2*.7854*B4)
12	Bore	4.125			
13					

Column A = input names
Column B = input values that you can change
Column C = output names
Column D = calculated results
Column E = actual formulas hidden in column D

Whenever you change values in Column B, the recalculated result displays in column D. The formulas in Column E are for reference only. They are actually hidden in the cells in Column D to make the calculations.

Now the piston speed for any stroke shown at B3 displays at D5 and it changes anytime you enter a new stroke dimension into B3. Cool, huh?

A more advanced problem would be to calculate an unknown bore size when all you know is the stroke and the displacement. Recall the formula from Chapter 1:

$$\text{Bore} = \sqrt{[\text{engine size or cid} \div (\text{stroke} \times 0.7854 \times \text{number of cylinders})]}$$

In this case: $\sqrt{[350 \div (3.48 \times 0.7854 \times 8)]}$

Step 1

To find this in the spreadsheet, type "Displ." into cell A8. That indicates you are going to enter a known displacement into cell B8.

Step 2

Now type "Bore?" into C8 to indicate that the calculated bore dimension displays in the adjacent cell D8.

Step 3

Type "Stroke" at A9 so you can enter a reference value at B9.

Step 4

Select D8 and enter the following formula into the formula bar. It is the spreadsheet version of the formula for finding the bore:

$$= \text{sqrt(B8/(B9*0.7854*B4))}$$

Step 5

Now you can enter a known engine size at B8 and it displays (in D8) the calculated bore size based on the other reference values.

You can arrange your input and output cells any way you want, as long as the appropriate cells are referred to in your formula entry. Note that (sqrt) in the bore formula tells it to take the square root of the references within the parentheses.

Squaring a number in a formula can be done in two ways. If you have 4.00 in cell B2, you can square it in the formula bar as follows:

$$\text{B2*B2 (which means 4.00 x 4.00)}$$

Or you can do it this way: B2^2 (which means $B2^2$).

	A	B	C	D	E	F
Formula:						
2	**Bore**	4	cid	349.85	B2*B2*B3*.7854*B4	
3	**Stroke**	3.48	cyl. vol.	43.73	B2*B2*B3*.7854	
4	**Cylinders**	8				
5	**RPM**	6000	P/S ft/min	3480	B3*B5/6	
6						
7						
8	**Displ.**	327	Bore?	4	sqrt(B6/(B7*.7854*B4))	
9	**Stroke**	3.25				
10						
11	**Displ.**	372	Stroke?	3.48	B11/(B12*B12*.7854*B4)	
12	**Bore**	4.125				
13						
14						
15						

Here's a screen shot of the same sample calculator built in Google's open document spreadsheet. It's a handy aid, but if your internet goes down you won't have access to it during the downtime.

The little arrow character pointing up tells the formula to raise the preceding value to the power immediately following it. It is called a caret, and it is the symbol for raising to a power. The number immediately following it indicates the required power. In this case the 2 after the caret tells the formula to raise the value in cell B2 to the second power. In other words, square the preceding value. If you were to put a 3 after the caret it would tell the formula to cube the B2 value, or the same as B2 x B2 x B2.

Finding Stroke Length

Suppose you have a displacement limit and you have decided to run the largest possible bore to help unshroud the intake valve and to run maximum piston area. Say the limit is 372.99 ci and you are planning a bore size of 4.125 inches. How do you set up the spreadsheet calculator to find the stroke that keeps you within your limit? First, ignore the 0.99 and just go with 372 to have some wiggle room to accommodate bearing tolerances and re-honing. Recall the formula from Chapter 1:

$$\text{Stroke} = \text{displacement} \div (\text{bore}^2 \times 0.7854 \times \text{number of cylinders})$$

$$\text{Stroke} = 372 \div (4.125^2 \times 0.7854 \times 8) = 3.479$$
$$\text{or } 3.48 \text{ inches}$$

All of the necessary data is present on the sample spreadsheet we have already constructed.

Step 1
Enter "stoke" at cell C11.

Step 2
To add a separate reference for this bore dimension, type "Bore" into A12 and use the input bore at B12 for your formula.

Step 3
Now select D11 and enter the following formula in the formula bar:

$$= B11/(B12\text{^}2*0.7854*B4)$$

Step 4
The result displays in D11. It tells you the maximum stroke you can run to stay legal.

This takes longer to explain than it does to actually do. The key is to remember that all of the formulas you enter have to refer to input cells that contain variable values that you can change. If the formula contains a mathematical constant, you can type it right into the formula. It doesn't have to reside in a reference cell even though you could do

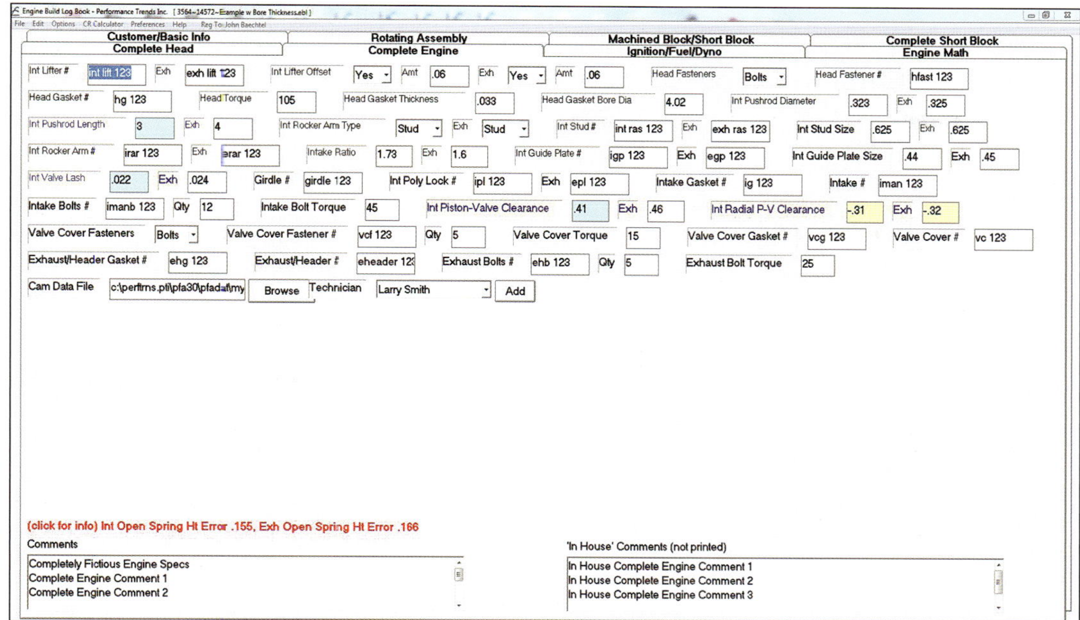

If you're not up to building your own engine content spreadsheet, you may like Performance Trends' Engine Log Book software that records all your part numbers and specs. It also performs checks and calculations on all your relevant inputs and organizes the results for handy professional-looking printouts.

it that way. On a hand-held calculator you can just enter 0.7854 and it knows what you mean.

Other Calculators

Refer to the symbols chart below to build the spreadsheet versions of the formulas presented in this book. The most useful way to apply these formulas is to build them into individual calculators based on the formulas found in each chapter. One way to do it is to group some of your favorite calculators on a single page and have the reference cells for all of them shown in columns A and B. Then you are always changing values in column B and reading the results elsewhere on the page within the individual calculators. That way, different calculators that sometimes require the same information all refer to the same input cell and you only have to input the value once.

You can also put different types of calculators on different pages. The following is a suggested list of the types of calculators you might want to create on different pages:

Page 1 displacement, compression ratios, and piston speeds
Page 2 horsepower, torque, and RPM formulas
Page 3 induction, cylinder heads, and exhaust formulas
Page 4 fuel systems, atmospherics, and combustion
Page 5 camshaft math
Page 6 other handy formulas

The following symbols will help you build your formulas in the formula bar:

=	formula follows
*	multiply
÷	divide
+	add
-	subtract
sqrt	take the square root
^	raise to the following power (e.g., $x^2 = x\char94 2$ and $x^3 = x\char94 3$)
( )	contains operations to be worked as a group

Once you get the hang of it, you'll find ways to incorporate all kinds of formulas besides those used for engine math. They might include gear ratios, trans ratios, wheel speeds, suspension travel, or whatever. Soon you'll be saving all your favorite formulas on your PC. Just don't forget to save your work each time you use them.

If you don't have Microsoft Excel you can purchase a copy online for a decent price. It's worth it because it can calculate any formula in this book and much more. It's available for PC or Mac and well worth the expense.

Eventually you will also figure out ways to build other calculators and even turn them into databases for parts and equipment—perhaps make your own engine build sheets with blank spaces for entering specs and clearances. The possibilities are endless and if you can't afford Excel, there are always open document versions on the Internet.

ENGINE BUILD/CONTENT SHEET

Component	Mfgr.	Part Number	Type	Size	Contact	Phone/e-mail

ENGINE ASSEMBLY SHEET

SPEC	Cyl. 1	Cyl. 2	Cyl. 3	Cyl. 4	Cyl. 5	Cyl. 6	Cyl.7	Cyl. 8
Bore								
Skirt								
Clearance								
Pin								
Pin Bore								
Clearance								
Mains								
Rods								
Deck Height								
Crank Thrust								
Cam End Play								
Top Ring Gap								
2nd Ring Gap								
Oil Ring Gap								
Rod Side Clearance								
Gasket Thickness								
Quench								
Int. Spring Height								
Exh. Spring Height								
Int. Valve to Piston								
Exh. Valve to Piston								
Intake Rocker Ratio								
Exh. Rocker Ratio								
Pushrod Length								
Lifter Bore								
Lifter								
Retainer to Seal								
Valve Lift								
Duration @ .050								

ENGINE SIMULATION AND MODELING SOFTWARE

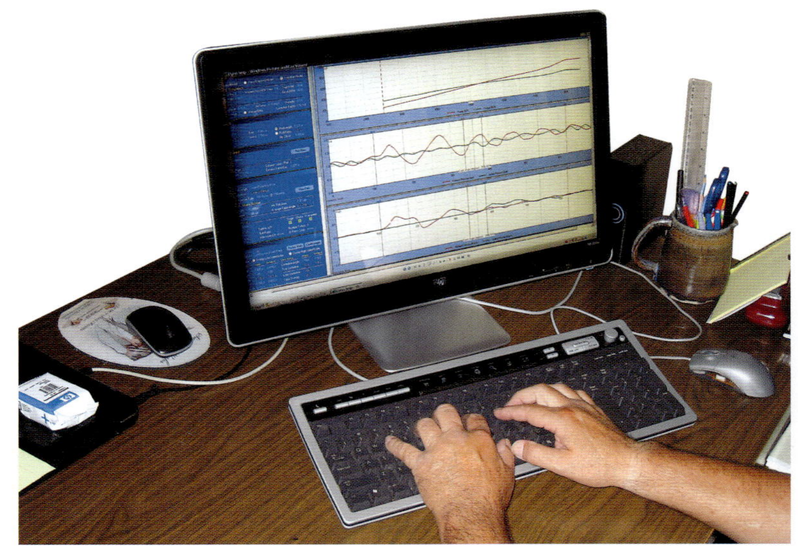

PC-based engine design and simulation software for home enthusiasts is affordable, surprisingly sophisticated, and capable of delivering accurate and highly instructive results.

Once you are comfortable with the formulas and concepts presented in this book, you can take advantage of more sophisticated tools that let you actually model and test your engine-building ideas on your home computer. Engine-simulation software has been around for more than 20 years, and it has steadily improved. Current programs are very robust and surprisingly affordable. Those presented here represent the cream of the crop and you can expect accurate and instructive results from all of them. We'll take a brief look at the top engine simulators first and then a number of support programs that allow you to test your engine models on the dragstrip and at Bonneville or any of the other top speed venues such as El Mirage and shorter one-mile tracks. In the end you will undoubtedly end up purchasing some of these programs

and you will thoroughly enjoy using them to test your ideas. They give you unparalleled freedom to design and test endless engine combinations while sitting around the house in your underwear and without spending a dime on dyno time.

Advanced computer skills are not required to operate these programs. They are all compatible with contemporary 32-bit operating systems up to and including Windows Vista and Windows 7. Most current home computers have the memory and processing speed to run them easily, although some advanced simulations may take a few minutes. If you bought your computer within the last decade you should have no trouble running these simulations. They walk you through all the steps, they provide excellent support and documentation and in

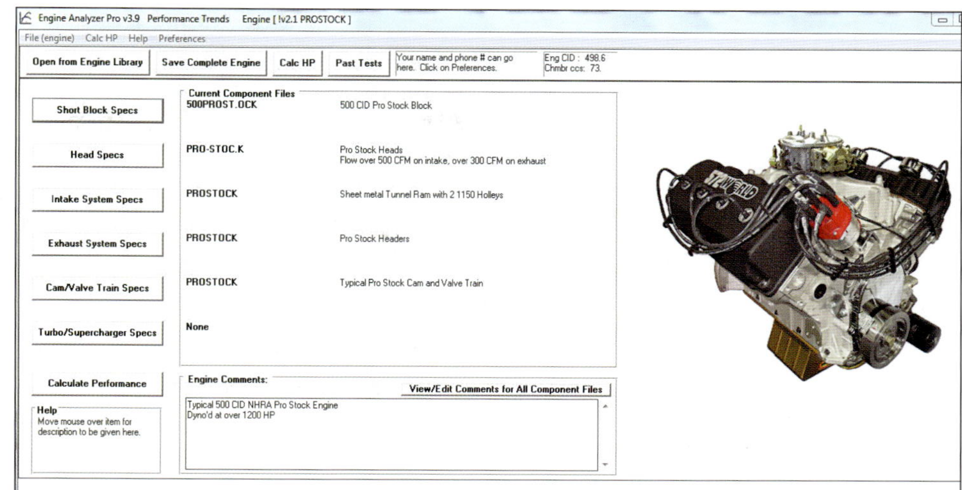

Performance Trends' Engine Analyzer series offers basic, mid-level, and high-end engine simulation for all types of engines including turbocharged, super-charged, and nitrous-oxide versions. It provides highly detailed results with graphic overlays for easy comparisons.

some cases, online updates. For most applications, you fill in the blanks with the appropriate specs and/or choose components from extensive menus. It's very easy and surprisingly instructive to the point that most programs will expand your knowledge and your engine planning skills just from using them.

Performance Trends

Performance Trends (www.performancetrends.com) offers three versions: Engine Analyzer, Engine Analyzer Plus, and Engine Analyzer Pro. All of them simulate almost any engine combination you can conceive, including bore and stroke sizes from 1 to 10 inches and up to 16 cylinders with multiple valves (2 to 6) and an RPM range from 500 to 30,000.

Engine Analyzer

The basic version runs simulations for gasoline and alcohol in normally aspirated mode or with turbocharged, supercharged, or nitrous options. It uses wave tuning algorithms for VE prediction and includes spark settings and detonation simulation. Supercharged applications can be modeled with and without intercooling and they include provisions for both centrifugal- and Roots-type superchargers. More than 70 engine specifications can be input to describe your proposed engine to the program. From this the simulation delivers up to 23 data outputs per RPM and 18 special calculations, such as displacement, valve flow areas, dynamic compression ratios, etc. The program estimates a cylinder head flow

curve generated from a percentage-based "flow efficiency" that you choose from a table, or you can input up to three flow data points from actual flow bench data.

The program includes utility calculation screens that help you model compression ratios and other auxiliary calculations needed to support the simulator. It plots torque and horsepower per RPM graphically with up to seven overlays for run-to-run comparisons. From input or selected cam data it shows valve lift per crankshaft degree to help you evaluate cam profiles. It uses ASCII (American Standard Code for Information Interchange)

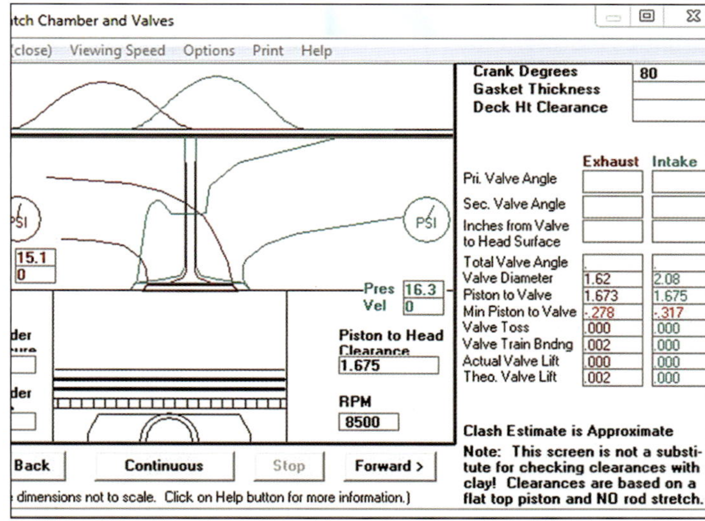

Final results in Engine Analyzer are displayed in tabular form and you can also display a cylinder simulation with a functional piston and Valves. You select the speed of the simulation and EA displays relative information such as piston to valve clearance, cylinder pressures and temperatures and corresponding crank angle.

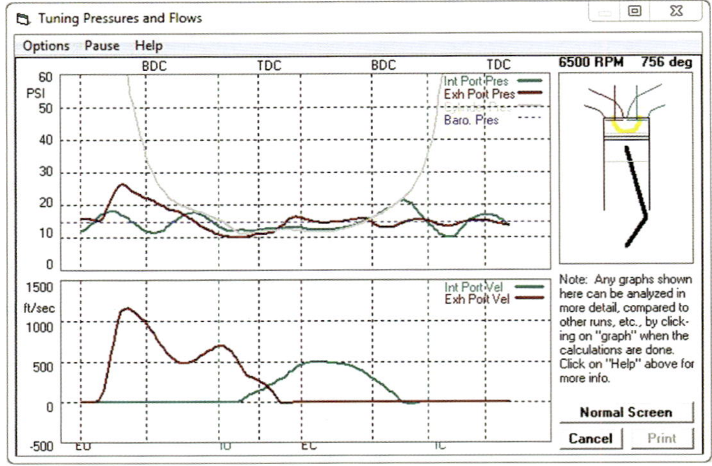

Engine Pro also incorporates intake port flow details relating valve lift to crank angle, piston speed and related flow demand. This screen displays a piston speed summary, shift points, and red line.

text so data can be written to a file with use-defined options for importing to spreadsheets such as Microsoft Excel. An optimize feature automatically chooses best port or runner size and length and can spec for best peak torque and horsepower or average torque and horsepower. You can input any 0.050-inch-lift cam value or a specified value including advertised timing plus lift, lash, rocker ratios, centerlines, ramp ratings, and so on.

Engine Analyzer Plus

The mid-level program has ten additional features, including diesel and alternate fuel modeling. Gasoline and alcohol can be modeled at standard and enriched air/fuel ratios for detonation control and you can also simulate with propane gas, propane LPG, or CNG (methane) fuels. The software adjusts spark for best power and predicts knock and retard timing to prevent detonation. This version incorporates 20 additional inputs including the ability to check valve-to-piston clearance and full cylinder head flow data. You can input flow volumes with up to 8 valve-lift increments. Each of the 23 outputs calculated for each RPM can also be graphed with 7 overlays. And you can substitute new engine mods right in the graph screen and immediately see changes to the power curve.

Engine Analyzer Pro

This version incorporates intake and exhaust port pressures based on Finite Difference wave simulation. It specifies net flow over the entire valve event including net trapped fresh mass charge and heat transfer to all chamber components. Simulations take longer in this version because of the greater number of calculations requested of the processor. Precise information calculated includes piston position, valve opening and flow, cylinder pressure, instantaneous torque, and piston thrust in one-tenth-degree increments up to 720 degrees to complete a full cycle. Successive cycles are repeated to obtain a constant result based on wave tuning that splits intake and exhaust runners into small sections to precisely calculate pressure, velocity, momentum, friction losses, heat transfer, and other pertinent information that can require up to 60 seconds or more of calculation time, depending on the degree of accuracy you have specified.

The program auto-selects best spark advance or you can input a forced spark curve with 6 break points that you can specify produces best power for the desired burn rate. It also calculates a knock index so you can model with trace detonation. While no program can fully predict conditions that may cause intermittent detonation in one or more cylinders, Engine Analyzer Pro delivers highly accurate predictions based on qualified assumptions.

This advanced version offers up to 50 outputs per RPM including some only applicable to special valvetrain dynamics or turbo/supercharger applications. With up to 42 special calculations, it includes cam specs at 0.200-inch tappet lift and lift at TDC. You can also specify metric inputs and outputs and graph 25 different types if you have cycle data including valve lift, piston thrust, port pressure, valve flow versus crank angle and cylinder volume (pressure/volume diagrams). In this version you can overlay up to 12 different tests for comparison purposes.

The "Chain Calculation" feature lets you increment one to six engine mods through two to six selected settings that may require several hours to calculate depending on your processor. Upon completion you can pinpoint the best power output and identify the components and optimum tuning specs that produced them.

Another unique feature is the ability to optimize for the Engine Masters Challenge requirements so you can predefine and model your proposed entry.

Engine Analyzer Pro is one of the most robust, affordable simulation programs available. All versions of the program create power curves that integrate into

Performance Trends' vehicle simulation programs such as Drag Race Analyzer, Circle Track Analyzer, or a fuel economy calculator. The best part is an auto-linking component that automatically provides an ET, 60-foot time, MPH, lap time, and miles per gallon right in the Engine Analyzer program for instant verification of results.

When visiting www.performancetrends.com, also check out the Rotating Inertia Calculator, which helps you predict the effects of an engine's rotating components on the overall "effective weight" of the vehicle.

Racing Systems Analysis

Racing Systems Analysis (www.quarterjr.com) offers a junior and a pro version of its engine simulator. Racing Systems Analysis (RSA) is best known for its Quarter Jr. drag racing analysis program that has been serving the drag racing community for more than 20 years. Quarter Jr. and Quarter Pro are two of the most popular drag racing simulators available for race modeling especially for those two have accurate dyno data on their engine combination. Engine Jr. and Engine Pro build on that solid reputation.

Engine Jr.

Engine Jr. is based on extensive collation of existing dyno test data matched with known engineering theory and fundamentals. It requires input of basic information including bore, stroke, rod length, compression ratio, cam specs, manifold type, valve specs, carb specs, carb type, and fuel. In return it provides peak horsepower, torque, power per cubic inch, and recommended redline and shift points. It also calculates minimum port cross-sectional area, mean flow velocity, and CFM per square inch of cross section.

Four auxiliary worksheets help you calculate compression ratio and estimate carb and throttle body flow, cylinder head flow, and minimum cross section at the intake valve throat. In addition to performance results it also provides a graphic dyno curve with tabular horsepower and torque in 100-rpm increments. It's a very powerful program for entry-level enthusiasts who want to model performance engine combinations.

Engine Pro

Engine Pro takes it to the next level with an intake port flow worksheet that accommodates bench data input. It helps you model intake and exhaust systems and allows you to examine intake ramming and wave tuning occurring in both tracts. It models port data and makes recommendations on valve size based on your inputs. It also provides extensive mechanical details regarding piston speed versus RPM and piston position versus crank angle. The intake flow detail screen uses your input data to determine piston flow demand relative to valve timing. You can also infer tips on how to obtain greater accuracy from your flow bench testing by using a higher test pressure than the standard 28 inches of water.

Once you have modeled your best engine combination, you can import test results to RSA's other performance software for drag racing and speed trails. These programs use your modeled data to predict track performance and they also accept actual dyno data if you have it. RSA also has an air density program (see Chapter 10) that helps with final tuning at the track.

Comp Cams/ProRacing Sim

Comp Cams is the parent company of ProRacing Sim. It's software is available via both of their Web sites: www.compcams.com and www.proracingsim.com.

DeskTop Dyno 5

This engine simulation software delivers extensive component testing at a high level of accuracy. It can model 4-cycle engines with up to 12 cylinders and displays horsepower, torque, VE, and other results on customizable full-color graphs and tables. ProRacing Sim claims accuracy to within 5 percent of actual dyno results. Extensive menus provide instant access to specs on thousands of engine components or you can input your own custom specs. The program features powerful auxiliary calculators for detailed analysis including an airflow pressure-drop calculator, induction flow calculator, and cam math calculator. Like most simulation programs, you can simply input your specs or select stored specs and watch the results materialize in seconds.

The "Quick Iterator" function provides automated testing that seeks optimum component combinations for any application. This handy tool selects components, performs virtual dyno tests, and pinpoints the ideal combinations. You can test engine speeds from 1,000 to

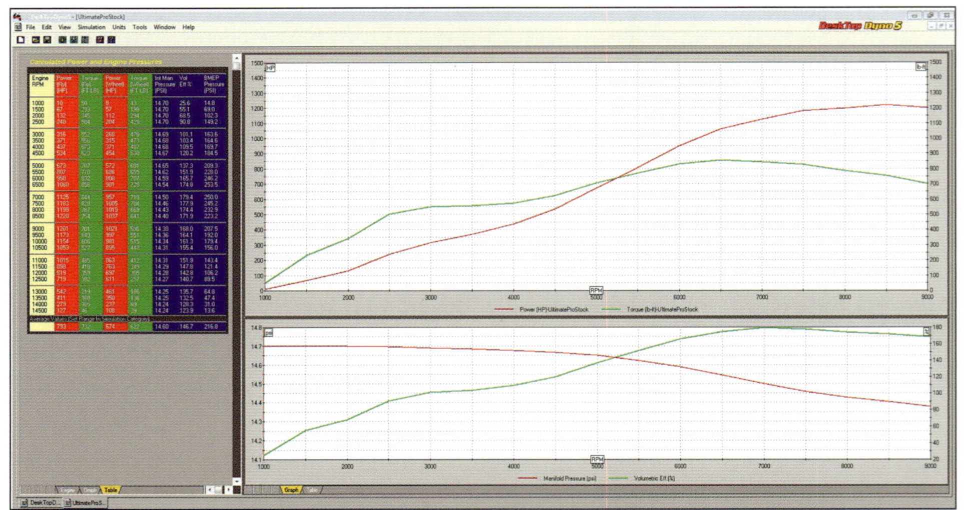

DeskTop Dyno's main input screen allows you to input all your engine specs and view power curves in adjacent graphs. It graphs horsepower, torque, VE, engine pressure, flow rates, and more.

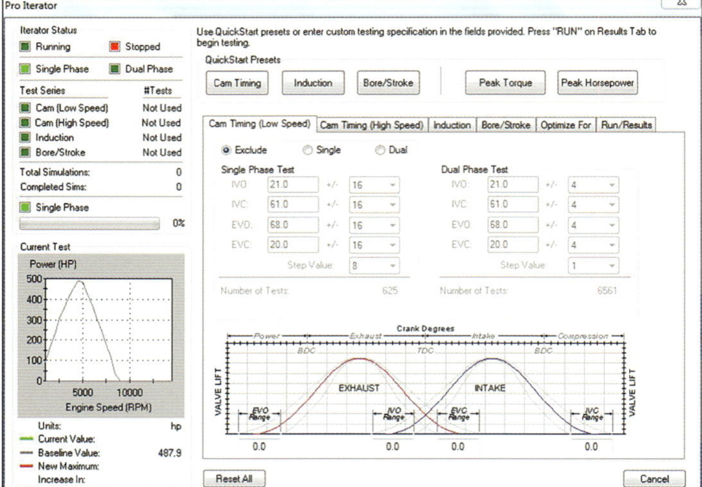

The Pro Iterator screen provides an automated testing regime that seeks optimum engine combinations with a single mouse click. It automatically dyno tests all possible combinations and displays the best performing results.

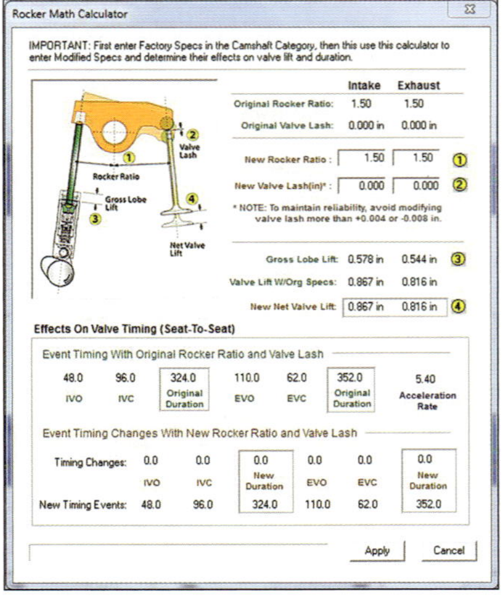

DynoSim5 features a rocker arm math calculator for viewing rocker ratios and their attending valve motion.

14,500 rpm and get results with scalable graphs of power, toque, and pressure curves. Then retest with alternate fuels or nitrous oxide if you wish. Test at any desired air/fuel ratio with carbs or injectors and any manifold type. The simulator models multiple valves and permits full manipulation of cam timing events. It also includes flow data models and cylinder heads with input specs. The built-in component library contains specs for almost every domestic and foreign engine including thousands of predefined short-block assemblies. It also comes with a 160-page full-color manual which you can download for free and study before you buy.

DynoSim5

Like DeskTop Dyno 5, DynoSim5 performs full-cycle simulations using the standard Filling and Emptying model. DynoSim5 delivers even more sophisticated simulations that include environmental effects, combustion and chamber shape modeling, ignition timing with advance curve graphing, new forced-induction modeling, a rocker ratio and lash calculator, and data files exportable to Microsoft Excel for further analysis. The program contains even more choices for multi-valve cylinder heads and it allows selection and evaluation of specific combustion chamber configurations.

Another unique feature native to DynoSim5 is the built-in compression math calculator that features two modes: a known volume mode and a burette measured mode. Either mode accepts user input of known or measured values and the burette measured mode provides alternate input windows to accept burette measurements

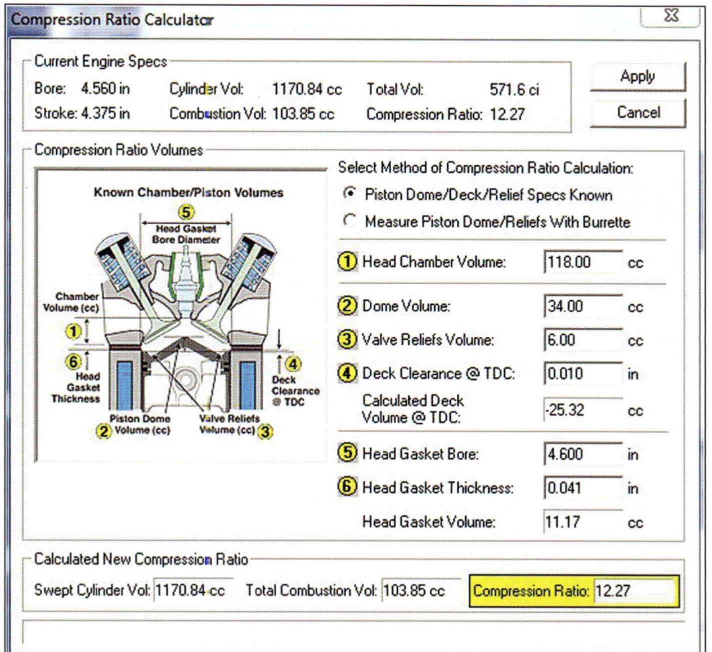

DynoSim5 features numerous auxiliary calculation screens including a compression-ratio math calculator that figures exact compression ratio from your input of component specs and measured volumes.

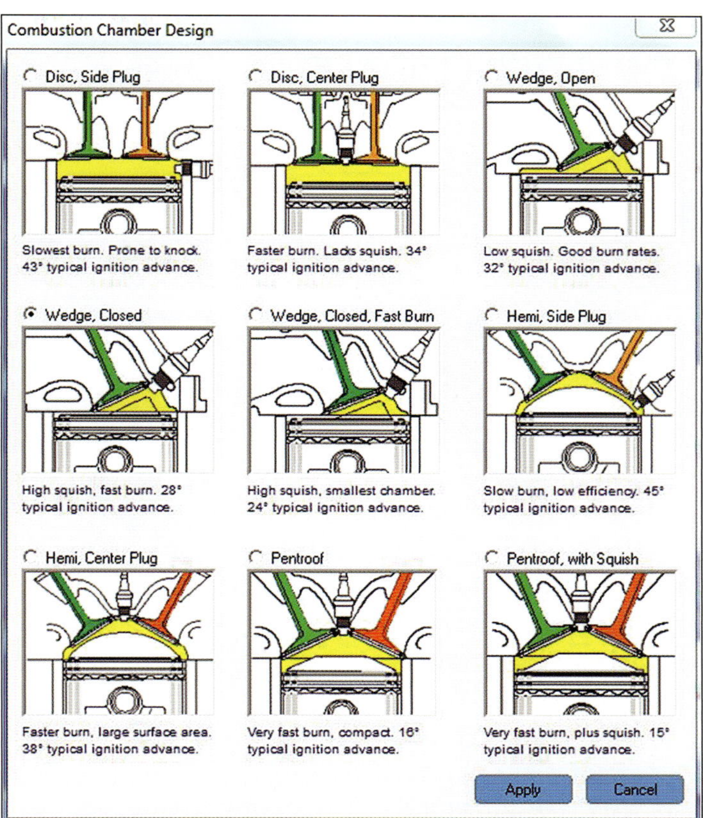

The combustion chamber selector allows you to specify your specific combustion chamber type and then models its characteristics including flow, burn rate, heat rejection to the cooling system, and other chamber components.

taken with the piston-down-the-bore method described in Chapter 3. The combustion chamber selector provides nine different combustion geometries based on chamber shape, burn rate characteristics and chamber timing requirements based on gasoline. DynoSim5 also incorporates an advanced ignition cycle model optimized for minimum advance best torque (MBT).

The amount of component selection and modeling in DynoSim5 is truly extraordinary. Be prepared to spend many hours simulating combinations while gaining a deeper insight into the dynamics of performance engines.

Motion Software

Motion Software is the originator of the DeskTop Dyno program. www.motionsoftware.com

Dynomation-5 Professional

Dynomation-5 Pro is the so-called Cadillac simulator based on extensive wave action simulation. It puts the user directly inside a running engine combination to view, analyze, and understand the powerful wave dynamics that influence induction and exhaust flow. It pro-

vides evaluation of specific intake runner lengths, section widths, port taper angles, header tubing and collector dimensions, and how they affect VE and the operational gas dynamics in the engine. Variations in pressure and flow velocities at the valve/chamber interface are displayed along with induction and exhaust pressure pulses and mass flow data.

The visual representation of these complex dynamics provides surprising insight into how torque and horsepower are manufactured inside a turning engine. A 3-D cutaway engine model shows mass flow, port velocities, and pressures all synchronized to crank angle for pinpoint analysis of camshaft profiles. It accepts test data from Cam Pro, Cam Pro Plus, S96, and Cam Doctor files. The 3-D model shows piston motion, gas dynamic flow, and visual representation of mass flow intake pressure waves and velocities and more throughout the entire four-cycle combustion process.

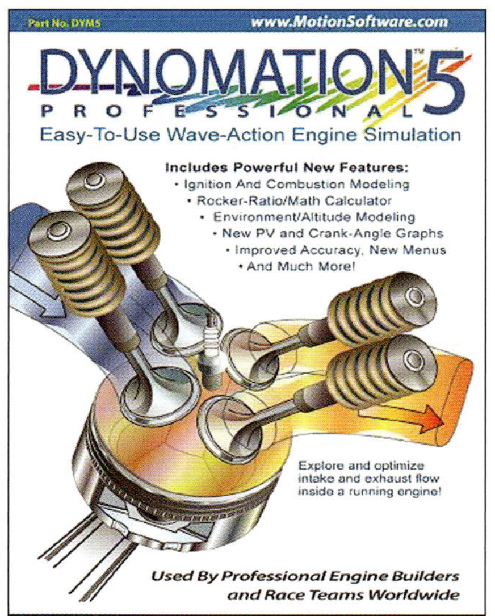

Dynomation 5 is a robust wave action simulator that has been very popular over the years.

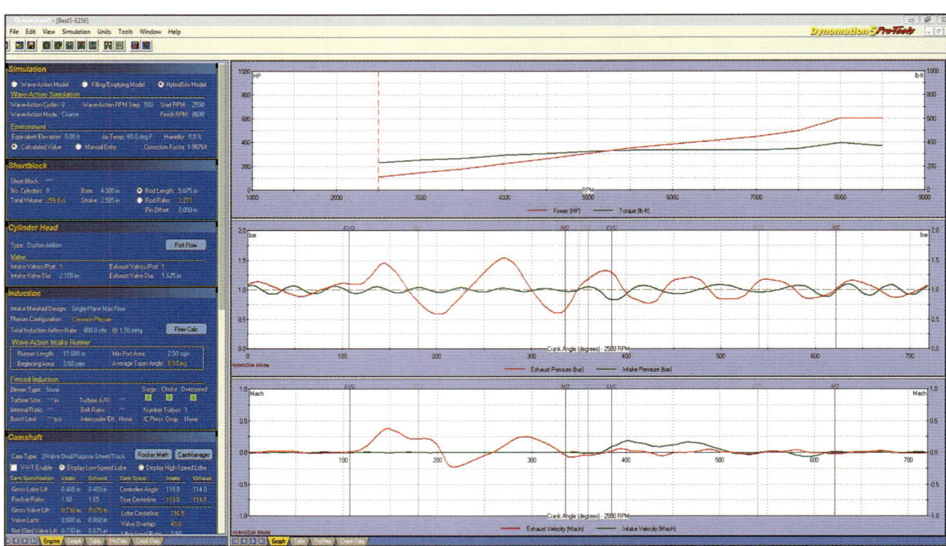

Dynomation 5's main screen displays wave simulated pressures based on components and specs that you specify. This program performs the wave modeling in real time while you observe it onscreen.

The seamless combination of the Filling and Emptying model and the wave dynamics in Method of Characteristics accurately predicts the interaction of complex pressure wave and gas particle flow. This pinpoints optimum port sizes, shapes, and lengths for both intake and exhaust plus integrated cam timing and valve motion analysis. Additional output data includes indicated horsepower, friction horsepower, pumping horsepower, mechanical efficiency, gas force on the piston, induction airflow, and piston speed. Pressure waves are shown on the 3-D engine, along with keys to the accompanying crank angle.

Dynomation also models most types of fuels and all boosted and nitrous-oxide applications with appropriate inputs for tight control of simulation characteristics. Full exhaust system modeling is also incorporated,

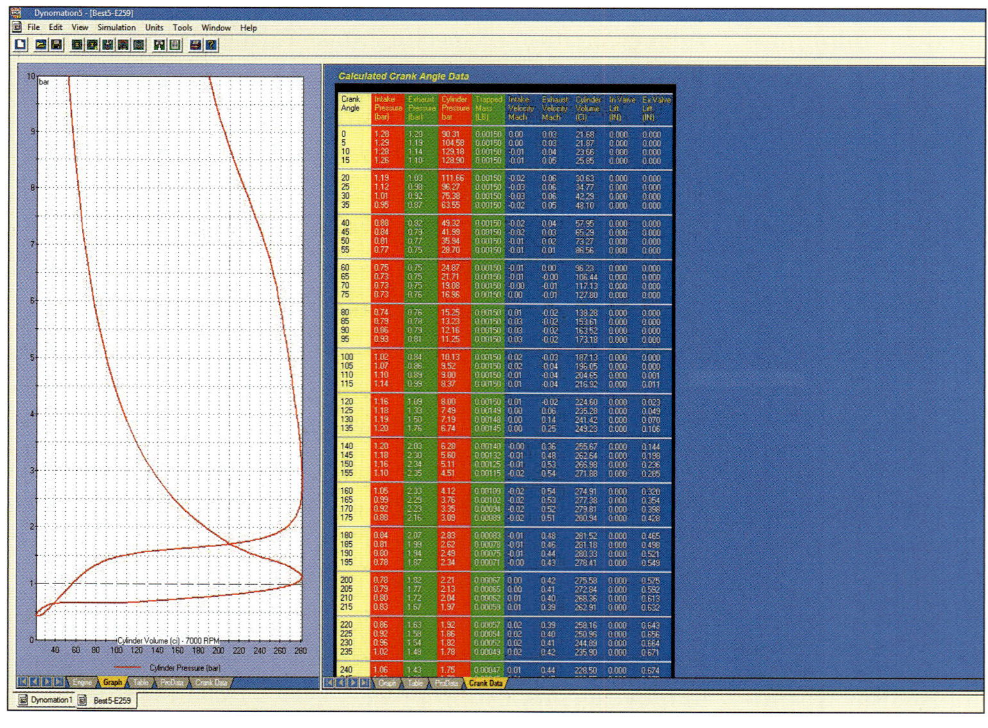

Pressure crank angle graphs and pressure volume diagrams show real time inlet, exhaust, and cylinder pressure changes and how they interact to affect power.

including mufflers, catalytic converter backpressure, and stepped headers plus collector size, length, and taper. If your engine runs variable valve timing, Dynomation can simulate it while providing full lobe acceleration modeling and analysis.

More than 6,000 real cam profiles are available for selection or you can input your own custom specs. Even professional engine builders do not underestimate the powerful insight that Dynomation provides.

Support Programs

Some of the most pertinent support programs include Racing Systems Analysis' DENSITY altitude program, Quarter Jr., Quarter Pro drag racing simulators, and Bonneville Pro, which uses real or simulated engine output data to model top speed efforts at Bonneville and other high-speed venues. RSA also offers *Engine Pro: The Book* with a free audio cassette for an in-depth tour of its engine simulation software.

Performance Trends backs up its Engine Analyzer series with multiple programs, including a comprehensive compression ratio calculator and fuel injector calculator that calculates injector size, pulse width and duty cycle, pressure corrections, EFI controllers, and other calculations unique to electronic fuel injection applications. They also offer Port Flow Analyzer for flow bench analysis and the Cam Analyzer for measuring and evaluating cam profiles.

And, finally, for the very hardcore among us, I recommend a visit to www.hi-techniques.com to check out the advanced engine analysis and data acquisitions components. While not for the home PC, its systems allow engine dyno operators to data log real-time cylinder pressure traces plotted against individual cylinder torque and horsepower output (see Chapter 5). This is the type of support that helps designers pinpoint deficiencies in various cylinders and redefine component relationships to bring individual cylinder power production in line with all contributing cylinders.

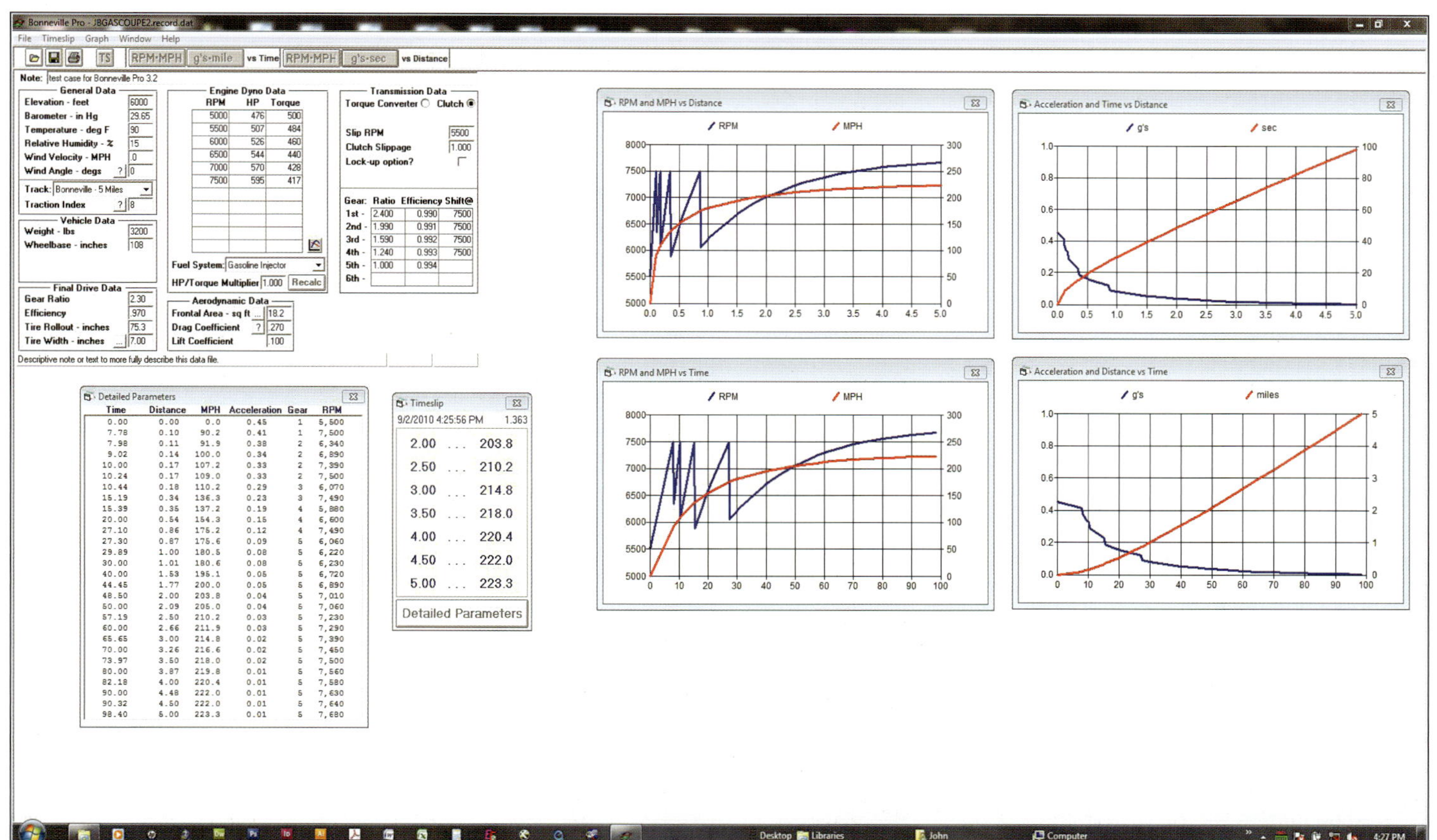

Bonneville Pro's input screen (top left) accepts real or modeled engine dyno data, vehicle aerodynamic data, track data, and weather conditions to predict performance as shown on the graphs (right) and MPH printouts (bottom left).

ABBREVIATIONS AND ACRONYMS

A/F: air/fuel ratio
ABDC: after bottom dead center
ATDC: after top dead center
A/R: area to radius ratio
bar: 14.68 psi or 1 atmosphere
BBDC: before bottom dead center
BTDC: before top dead center
BS: bore spacing
BSFC: brake specific fuel consumption
BMEP: brake mean effective pressure
BDC: bottom dead center
cid: cubic inch displacement
ci: cubic inch
c/s: cross-sectional area
cc: cubic centimeter
CR: compression ratio
cfm: cubic feet per minute
cc/min: cc per minute
DA: density altitude
DR: displacement ratio
displ.: displacement
dr: density ratio
E/I: exhaust-to-intake ratio
EO: exhaust opening
EC: exhaust closing
ECR: effective compression ratio
EFI: electronic fuel injection
FMEP: friction mean effective pressure
ft/min: feet per minute
ft/sec: feet per second
g/sec: grams per second
Gen 1: First-generation Chevy small-block
Gen II: Second-generation Chevy small-block (LT1)
Gen III: Third-generation Chevy small-block (LS series)
gal/hr: gallons per hour
HP: horsepower
Hg: mercury
IMEP: indicated mean effective pressure

IC: intake closing
IO: intake opening
LSA: lobe separation angle
L/D: lift-to-diameter ratio
L: liter
lbs/min: pounds per minute
lbs/hr: pounds per hour
MEP: mean effective pressure (generic for cylinder pressure)
mbt: maximum brake torque
mm: millimeter
ME: mechanical efficiency
MAP: manifold air pressure
MAF: mass air flow
MAT: manifold air temperature
OEM: original equipment manufacturer
O_2: oxygen sensor
psia: absolute pressure
psig: gauge pressure
pi (π): 3.1417
PLAN: cylinder pressure, stroke length, piston area, RPM
PR: pressure ratio
RR: rocker ratio
RPM: revolutions per minute (engine speed)
R/S or R/L: rod length to stroke ratio
TQ: torque
TDC: top dead center
STP: standard temperature and pressure
SAE: Society of Automotive Engineers
WOT: wide open throttle
λ: lambda
V_1: cylinder volume
V_2: components of CR formula (total combustion space)
$\therefore$: therefore
$\approx$: approximately
X^2: units squared, or X times X
X^3: units cubed, or X times X times X
$\sqrt{X}$: square root of X
$\sqrt[3]{X}$: cube root of X

HANDY CONVERSION FACTORS

Convert From	Multiply By
atmospheres to bars	1.01325
atmospheres to inches of mercury	29.3921256
atmospheres to inches of water	406.80172
atmospheres to pounds per square inch	14.695649
bars to atmospheres	0.9869233
bars to inches of mercury	29.529983
bars to inches of water	401.48713
bars to kilopascals	10.0
bars to millibars	100.0
bars to pounds per square foot	2088.5434
bars to pounds per square inch	14.503774
centimeters to feet	0.0328084
centimeters to inches	0.3937008
centimeters to millimeters	1.0
centimeters per second to feet per second	0.0328084
cubic centimeters to cubic inches	0.0610237
cubic centimeters to liters	0.001
cubic centimeters to milliliters	1.0
cubic centimeters to ounces	0.03381
cubic feet to cubic inches	1728.0
cubic inches to cubic centimeters	16.387064
cubic inches to cubic feet	0.0005787
cubic inches to liters	0.0163871
feet per second to centimeters per second	3.48
feet per second to feet per minute	6.0
feet per second to miles per hour	0.6818182
foot pounds to Newton-meters	1.35582
gallons to cubic centimeters	3785.4118
gallons to cubic inches	231.0
gallons to liters	3.7854118
gallons, ethanol to pounds	7.6
gallons, gasoline to pounds	6.0
gallons, methanol to pounds	6.7
gallons, nitro methane to pounds	9.4
gallon, water to pounds	8.3449037
grams to milligrams	100.0
grams to ounces	0.0356274
grams to pounds	0.0022046
horsepower to pounds-feet per minute	3,300.0
horsepower to pounds feet per second	55.0
horsepower, metric to SAE	0.9863201
horsepower, SAE to horsepower, metric	1.0138697
inches to centimeters	2.54

Convert From	Multiply By
inches to feet	0.0833333
inches to millimeters	25.4
inches to mils	100.0
inches of mercury to atmospheres	0.03347211
inches of mercury to bars	0.0338639
inches of mercury to inches of water	13.595913
inches of mercury to kilopascals	3.3863886
inches of mercury to millibars	33.863886
inches of mercury to pounds per square foot	7.726197
inches of mercury to pounds per square inch	0.4911541
inches of water to atmospheres	0.0024582
inches of water to bars	0.0024907
inches of water to inches of mercury	0.0735515
inches of water to pounds per square inch	0.0361251
kilopascals to atmospheres	0.0098692
kilopascals to bars	0.01
kilopascals to inches of mercury	0.2952998
kilopascals to inches of water	4.0148716
kilopascals to pascals	0.001
kilopascals to pounds per square inch	0.1450377
liters to cubic centimeters	100.0
liters to cubic inches	61.023744
liters to gallons, USA	0.2641721
millimeters to centimeters	0.1
millimeters to inches	0.0393701
Newton-meters to foot pounds	0.737562
pascals to bars	0.00001
pascals to inches of mercury	0.0002953
pascals to inches of water	0.0040149
pascals to kilopascals	100.0
pascals to pounds per square foot	208.85434
pounds per square inch to atmospheres	0.068046
pounds per square inch to bars	0.0689476
pounds per square inch to inches of mercury	2.0360207
pounds per square inch to inches of water	27.681566
pounds per square inch to pounds per foot	144.0
pounds-feet to pounds-inches	12.0
pounds-inches to pounds-feet	0.0833333
square centimeters to square inches	0.01550003
square feet to square inches	144.0
square inches to square centimeters	6.4516
square inches to square feet	0.0069444

GLOSSARY

Absolute Pressure
Atmospheric pressure at sea level (14.7 psi) with no other pressure applied. An absolute pressure gauge (psia for psi absolute) reads the ambient pressure at the location where the measurement is being taken. If you apply 5 pounds of boost to an absolute pressure gauge at sea level it reads 19.7 psi. In contrast, gauge pressure is a direct pressure reading on a gauge that reads zero at any local pressure until you apply additional pressure boost. It is noted as psig for psi gauge. Most automotive gauges read gauge pressure.

Acceleration
The rate of increase in speed or velocity of an object.

After Bottom Dead Center (ABDC)
Position where the piston is accelerating away from bottom dead center (BDC). Intake valve closing occurs in this area.

After Top Dead Center (ATDC)
Piston position where the piston has passed top dead center (TDC) and is accelerating away from it. The exhaust closing event usually occurs in this area.

Air Density
The mass per unit volume of the earth's atmosphere at a given location. Air density is inversely proportional to altitude, or density decreases with increasing altitude. Therefore air density is reduced at higher elevations.

Air/Fuel Ratio
The ratio of air mixed with fuel for the purpose of combustion in an internal combustion engine. The ideal ratio for best power usually falls between 12.8:1 and 13.5:1. That is 12.8 parts air to 1 part fuel. With excess fuel, the mixture is rich (12.1:1 or less). If there is excess air, the mixture is lean (15.1:1 or higher). (See stoichiometric ratio.)

A/R Ratio
The square area of the turbine housing throat at the tongue divided by the radius from the axis of turbine wheel rotation to the centroid or dynamic center of flow through the volute.

Ambient Conditions
This refers to current local conditions in terms of temperature and pressure. A sample of ambient conditions for any given day might include the current temperature, atmospheric pressure (barometer), humidity, and water vapor pressure.

Atmospheric Pressure
14.68 pounds per square inch at sea level with a barometer of 29.92 Hg and 60-degree air with no humidity.

Base Circle
The round portion of a camshaft lobe where no motion is imparted to the valve lifter. When a lifter is on the base circle of the cam, there is zero lift.

Before Bottom Dead Center (BBDC)
Piston position before reaching bottom dead center. Normal area for exhaust opening event.

Before Top Dead Center (BTDC)
Piston position (typically number-1) before or approaching top dead center. The piston is decelerating and the intake valve opening normally occurs in this area.

Bob Weight
An adjustable weight temporarily attached to a crankshaft to simulate the mass of the piston, rings, rod, and bearings for the purpose of balancing the crankshaft.

Boost and Bars
Boost is the increase in manifold pressure supplied by either a supercharger or a turbocharger. One bar of boost is equal to 14.504 psi or 1 atmosphere. A 1-bar MAP sensor reads vacuum only. A 2-bar MAP sensor reads pressure above atmospheric up to 15 psi. A 3-bar sensor reads up to 30 psi above atmospheric. These are all absolute pressure sensors. Hence, a 3-bar sensor reads ambient atmospheric pressure plus up to 30-psi boost. You need a 2- or 3-bar MAP sensor to read boost pressure.

Bore
The circular diameter of an engine's cylinders.

Bore Spacing
The distance between centerlines of adjacent cylinder bores.

Bore/Stroke Ratio
An engine's bore dimension divided by its stroke length. If the bore is larger the engine is said to be over-square. If the bore is smaller, the engine is under-square. Square and over-square engines are typically more efficient due to superior breathing characteristics.

Bottom Dead Center (BDC)
The point where a piston is at the absolute bottom of its travel in a cylinder bore.

Brake Horsepower (bhp)
Horsepower as calculated from observed torque on an engine dyno.

Brake Mean Effective Pressure (BMEP)
Mean cylinder pressure calculated from observed torque on an engine dyno.

Brake Specific Fuel Consumption (BSFC)
Pounds of fuel burned per hour divided by horsepower. It is not an indicator of rich or lean, but rather a measure of engine efficiency. Highly efficient performance and racing engines often generate a BSFC of 0.35 to 0.42, much less than 1/2 pound of fuel per horsepower per hour. Production engines are now approaching these values as automakers improve efficiency. The general rule for sizing fuel injectors is 0.50 lbs/hr for gasoline. Most super-charged and turbocharged engines require 0.55 to 0.60 for best performance under boost.

Burn Rate
The flame propagation speed of the combustion process. It is determined by the hydrocarbon components in the fuel and is not related to octane which is a measure of resistance to detonation. Fast burning fuels are required for best power in modern high speed engines.

Cam Lobe
The eccentric portion of a camshaft that opens the valves by raising the valve lifters. In essence, the cam lobe is a device that converts rotary motion into linear motion.

cc
Cubic centimeters (cm^3). Typically a metric measurement of engine displacement, combustion chamber size, or fuel injector capacity.

Centroid
The dynamic center of exhaust gas flow through the volute of the turbine housing. Not usually the dimensional center, but typically biased toward the outer half. The point where half of the gas flow is above the center and half of it is below the center.

Charge
A reference to the fuel/air mixture that enters the cylinder for combustion, i.e., the incoming charge.

Charge Speed
The speed of the intake charge in feet per second. Also known as port velocity.

Center-to-Center Length
The distance between the centerline of the piston pin bore and the crank pin bore of a connecting rod.

Compression Ratio
The ratio between a cylinder's swept volume at BDC and its final combustion volume when the piston is at TDC. Typical compression ratios range from 8:1 to 11:1.

Combustion Chamber
The combustion space in a cylinder head. For a gasoline engine, the combustion chamber has at least one intake valve, one exhaust valve, and one spark plug. Four-valve and dual-plug chambers are used in some high-performance engines to achieve greater efficiency.

Compression Height
Distance from the centerline of the piston pin to the top of the piston. Same as pin height.

Compression Stroke
The second of four strokes in the internal combustion cycle. After the intake stroke, this stroke compresses the air/fuel mixture (charge) to increase its density prior to the spark that initiates combustion for the power stroke.

Compression Test
A test that measures cylinder pressure in a cranking engine. It looks for consistency of piston ring and valve seal from cylinder to cylinder; not necessarily the highest number, but one within a specified range.

Compressor Map
A graphic representation of a turbocharger's overall efficiency based on pounds per hour of fuel used plotted against pressure and density ratios.

Constant
A mathematical component or expression that maintains a permanent value for the purpose of computation. For example, in the torque and horsepower formula, 5,252 is the constant that enables the calculation of torque, horsepower, or engine speed if any two values are known.

Crank Angle
The crankshaft angle in relation to the piston top position (or pin centerline) in the bore for any given point in the stroke.

Crank Pin
The connecting rod throw on a crankshaft. This is typically called the rod journal.

Crevice Volume
The small open space between the piston and the cylinder wall above the top piston ring.

Cubic Inch
A three-dimensional space whose length, width, and height of 1 inch define its volume. One cubic inch equals 16.387 cubic centimeters and 1 cubic foot equals 1,728 cubic inches.

Cylinder Head
A separate engine component containing the combustion chambers, valves, spark plugs, intake and exhaust ports, and valve gear.

Deck Height
The distance the piston top is above or below the deck surface of the cylinder block when the piston is at TDC. If the piston top is below the deck surface it is called positive deck height. If it is above the deck surface it is called negative deck height.

Degree Wheel
A graduated (in degrees) aluminum disk or wheel temporarily attached to the front of a crankshaft to indicate crank position (in degrees) relative to piston position and valve event.

Density Altitude
The density of local air compared to known altitude conditions for standard atmospheric conditions based on air temperature, pressure, and water content.

Density Ratio
The ratio of oxygen content or air density of compressed air relative to the same volume at ambient pressure.

Detonation
The auto-ignition of the fuel mixture prior to the spark event. Caused by high cylinder pressure, unstable mixture quality, excessive localized heat, or a combination thereof. Often called pinging due to the annoying sound.

Displacement
The swept volume of an engine's cylinders measured in cubic inches, cubic centimeters, or liters.

Displacement Ratio
Cylinder volume divided by combustion space volume. It is always 1 less than the compression ratio. Most often used to help calculate the correct amount of cylinder head milling to increase compression ratio.

Dome Volume
The volume in cubic centimeters displaced by a dome on top of the piston. The dome is used to raise the compression ratio by reducing the combustion space above the piston.

Dyno
A water brake device that measures engine torque via a strain gauge.

Exhaust Closing Point (EC)
Measured in degrees after top dead center (ATDC).

Effective Compression Ratio
The combination of static compression ratio plus boost pressure: boost divided by ambient pressure plus 1 and multiplied by the static compression ratio. It is used to calculate the engine's compression ratio tolerance for pump gasoline.

Efficiency Island
The irregular shapes of various efficiency levels on a turbocharger compressor map.

Engine Cycle Analysis (ECA)
A SuperFlow Technologies Group software analysis program that utilizes a rotary shaft encoder unit on the crankshaft and a pressure transducer in the combustion chamber to directly measure real-time cylinder pressure relative to crank position on a running dyno engine.

Exhaust Opening Point (EO)
Measured in degrees before bottom dead center (BBDC).

Equivalent Displacement
A calculation that relates the displacement of a rotary engine to that of a reciprocating engine. ED = SV x 3, where SV is the swept volume of one rotary chamber.

Flow Area
The available space through which the air/fuel mixture or exhaust by-products enter and leave the engine's combustion space. This area is typically restricted at the valve throat area or at some point farther up the port.

Foot-pound
A measure of force times distance, or "work." If you move 10 pounds a distance of 50 feet you have performed 500 ft-lbs of work. If you do it over time, say 30 seconds, you have produced 16.66 ft-lbs/sec of power (500/30 = 16.66). Divide that by 550 (1 hp = 550 ft-lbs/sec) and you have applied 0.030 hp.

Friction Horsepower
Friction horsepower is the total power absorbed by the mechanical components of the engine due to frictional forces that oppose motion by converting it to heat.

Friction Mean Effective Pressure
This is the loss incurred by friction between the pistons and rings against the cylinder walls and the bearings on the crankshaft and rods. It is a calculated number.

Gauge Pressure - See absolute pressure

Intake Closing Point (IC) - Measured in degrees ABDC.

Indicated Horsepower
The maximum power a given engine can theoretically produce.

Indicated Mean Effective Pressure
The force generated against the piston top by the combustion process. It is directly measured by a pressure transducer in the chamber.

Injector Pulse
The time in milliseconds that a fuel injector is held open to flow a given amount of fuel.

Injector Size
An electronic fuel injector's flow rating in lbs/hr at 85-percent duty cycle. Also noted in cc/sec for most import applications.

Installed Height
The installed valvespring height when the valve is closed.

Intake Valve Opening Point (IO)
Normally measured in degrees before top dead center (BTDC).

Lobe Displacement Angle (LDA)
The number of camshaft degrees separating the centerlines of the intake and exhaust lobes. Also referred to as Lobe Separation Angle (LSA) or simply "Lobe Center."

Lobe Separation Angle (LSA)
Same as Lobe Displacement Angle. Most commonly used reference.

Mean Effective Pressure (MEP)
The combustion pressure above the piston on the power stroke.

Overbalancing
The practice of adding a small percentage more than the normal weight of 50 percent to the bob weights when balancing an engine. It helps smooth the engine's operation at high RPM.

Overlap
A brief period in camshaft timing where both valves are partially open at the same time.

Pin Height
Distance from the centerline of the piston pin to the top of the piston. Formally known as the compression height.

Power Curve
A graphic representation of an engine's torque and horsepower output per RPM. On a graph, the torque and the horsepower curves cross at 5,252 rpm.

Pressure Ratio
The ratio of artificial boost pressure to ambient atmospheric pressure. Calculated by adding boost pressure and ambient pressure then dividing by ambient pressure.

Reversion
The consequential movement of air/fuel mixture or residual exhaust products back into the port from where it came as a result of piston motion, or a pressure differential caused by valve opening and closing events during overlap when both valves are open at the same time.

Rocker Arm Ratio
A ratio by which the rocker arm multiplies camshaft lift to increase total valve lift.

Rotating Weight
The amount of rotating crankshaft weight that resists rotating motion.

Saturation Point
The point at which the intake valve curtain area or flow window matches or exceeds the smallest flow area in the runner, causing the port to become the restriction instead of the valve.

Shrouding
A design configuration where inlet and/or exhaust flow is partially restricted by the proximity of an adjacent combustion chamber wall or the actual cylinder wall. It affects the efficient flow of gases in and out of the combustion chamber.

Siamesed Bore
A block casting configuration where solid material is cast between the cylinders with no space for a cooling jacket. A design intended to increase cylinder wall strength.

Standard Temperature and Pressure (STP)
Atmospheric conditions of 29.92 Hg barometer with 60-degree (F) dry air. This is the standard correction factor used for most performance industry and magazine dyno testing. It results in corrected numbers approximately 4 percent higher than the SAE standard.

Supercharger
A belt-driven device that forces air into an engine to increase its breathing capacity and horsepower output.

Swept Volume
The volumetric space swept by a piston top as it travels up and down in the bore.

Tongue
The beginning of the inner opening in the volute where incoming exhaust gasses are first exposed to the turbine wheel.

Top Dead Center (TDC)
Position where the piston is at the exact top of its travel in the cylinder. It splits the dwell point where the rod changes its angle and represents the zero degree reference point for crankshaft degrees and cam timing events.

Torque
A twisting force expressed in ft-lbs or in Newton-meters for metric applications. One ft-lb equals 1.35582 Nm and 1 Nm equals 0.737562 ft-lb.

Torque Plate
A thick metal bar with properly spaced bore holes that is bolted to the cylinder block deck surface to simulate the distortion of cylinder head bolts during cylinder honing.

Turbine Housing Area
The area in inches of the turbine housing throat at the end of the tongue perpendicular to the direction of exhaust gas flow.

Turbocharger
An exhaust driven air compressor device that forces air into an engine to increase VE and thus horsepower output.

Valve Curtain Area
The circular flow window that opens when an intake valve is lifted off its seat by the camshaft.

Valve Lift
Distance the valve is lifted off its seat by the camshaft and valvetrain.

Valvetrain Geometry
A precise relationship between the rocker arm tip and the valve stem that provides optimum lift with minimal side thrusting and wear.

Variable
A mathematical component that can assume any assigned value for the purpose of computation.

Volumetric Efficiency (VE)
The ratio of an engine's actual airflow to its potential or calculated air capacity.

Volute
The scroll-shaped passage in the cast iron turbine wheel housing. It begins at the flange or foot and follows a decreasing circular path to the turbine wheel.

SOURCE GUIDE

Airflow Research
28611 W. Industry Drive
Valencia, CA 91355
877/892-8844
www.airflowresearch.com

Accessible Technologies, Inc.
(ATI) ProCharger
14801 W. 114th Terrace
Lenexa, KS 66215
913/338-2886
www.ProCharger.com

Automotive Racing Products
531 Spectrum Circle
Oxnard, CA 93030
800/826-3045
805/339-2200
www.arp-bolts.com

Barry Grant Fuel Systems
1450 McDonald Road
Dahlonega, GA 30533
706/864-8544
www.barrygrant.com

Comp Cams
3406 Democrat Rd.
Memphis, TN 38118
901/795-2400
800/999-0853 Cam Help
www.compcams.com

Drag Racing Pro
www.dragracingpro.com

Eagle Specialty Products
8530 Aaron Lane
Southaven, MS 38671
662/796-7373
www.eaglerods.com

Ford Racing Performance Parts
44050 N. Groesbeck Highway
Clinton Township, MI 48036-1108
586/468-1356 Tech
www.fordracing.com

Hi-Techniques, Inc.
2515 Frazier Avenue
Madison, WI 53713
800/248-1633
www.hi-techniques.com

Holley Performance Products
1801 Russellville Road
Bowling Green, KY 42101
270/781-9741
www.holley.com

icengineworks
www.icengineworks.com
800/597-3312

Jon Kaase Racing Engines
735 W. Winder Ind. Parkway
Winder, GA 30680
770/307-0241
www.jonkaaseracingengines.com

Motion Software
222 South Raspberry Lane
Anaheim, CA 92808-2268
714/231-3801
Motionsoftware.com

NHRA
2035 Financial Way
Glendora, CA 91740
818/914-4761
www.nhra.com

Performance Trends, Inc.
Box 530164
Livonia, MI 48153
248/473-9230
www.performancetrends.com

Pipe Max
Meaux Racing Heads
9827 Hwy 343
Abbeville, LA 70510
337/652-6220
maxracesoftware@yahoo.com

Powerbuilt
1431 Via Plata
Long Beach, CA 90810-1462
800/423-3598
www.alltradetools.com

Powerhouse Products
3402 Democrat Road
Memphis, TN 38118
800/872-7223 orders
901/795-7600
www.powerhouseproducts.com

ProRacing Sim
535 W. Lambert, Bldg. E
Brea, CA 92821-3911
714/255-2931
901/259-2355
www.proracingsim.com

Reher & Morrison Racing Engines
1120 Enterprise Place
Arlington, TX 76001
817/467-7171
www.rehermorrison.com

Racing Systems Analysis
www.RacingSecrets.com

Speed-Pro
Sealed Power Corporation
100 Terrace Plaza
Muskegon, MI 49443
616/724-5200
www.federal-mogul.com

Summit Racing Equipment
P.O. Box 909
Akron, OH 44309-0909
800/230-3030 Orders
330/630-0230 Customer Service
www.summitracing.com

SuperFlow Technologies Group
3512 North Tejon
Colorado Springs, CO 80907-5299
719/47-1746
www.superflow.com

Turbonetics
2255 Agate Court
Simi Valley, CA 93065
805/581-0333
www.turboneticsinc.com

Weiand
1801 Russellville Road
Bowling Green, KY 42101
270/781-9741   Tech Line

World Products
35 Trade Zone Drive
Ronkonkoma, NY 11779
631/981-1918
www.worldcastings.com